Jan Rhensius

Spin Dynamics and Spin Configuration in Nanopatterned Elements

AF004738

Jan Rhensius

Spin Dynamics and Spin Configuration in Nanopatterned Elements

Experimental results on patterned ferromangetic half-metals and field- & current-induced dynamics in Permalloy

Südwestdeutscher Verlag für Hochschulschriften

Impressum/Imprint (nur für Deutschland/only for Germany)
Bibliografische Information der Deutschen Nationalbibliothek: Die Deutsche Nationalbibliothek verzeichnet diese Publikation in der Deutschen Nationalbibliografie; detaillierte bibliografische Daten sind im Internet über http://dnb.d-nb.de abrufbar.
Alle in diesem Buch genannten Marken und Produktnamen unterliegen warenzeichen-, marken- oder patentrechtlichem Schutz bzw. sind Warenzeichen oder eingetragene Warenzeichen der jeweiligen Inhaber. Die Wiedergabe von Marken, Produktnamen, Gebrauchsnamen, Handelsnamen, Warenbezeichnungen u.s.w. in diesem Werk berechtigt auch ohne besondere Kennzeichnung nicht zu der Annahme, dass solche Namen im Sinne der Warenzeichen- und Markenschutzgesetzgebung als frei zu betrachten wären und daher von jedermann benutzt werden dürften.

Coverbild: www.ingimage.com

Verlag: Südwestdeutscher Verlag für Hochschulschriften GmbH & Co. KG
Heinrich-Böcking-Str. 6-8, 66121 Saarbrücken, Deutschland
Telefon +49 681 37 20 271-1, Telefax +49 681 37 20 271-0
Email: info@svh-verlag.de

Approved by: Konstanz, Universität Konstanz, Diss., 2011

Herstellung in Deutschland:
Schaltungsdienst Lange o.H.G., Berlin
Books on Demand GmbH, Norderstedt
Reha GmbH, Saarbrücken
Amazon Distribution GmbH, Leipzig
ISBN: 978-3-8381-0823-0

Imprint (only for USA, GB)
Bibliographic information published by the Deutsche Nationalbibliothek: The Deutsche Nationalbibliothek lists this publication in the Deutsche Nationalbibliografie; detailed bibliographic data are available in the Internet at http://dnb.d-nb.de.
Any brand names and product names mentioned in this book are subject to trademark, brand or patent protection and are trademarks or registered trademarks of their respective holders. The use of brand names, product names, common names, trade names, product descriptions etc. even without a particular marking in this works is in no way to be construed to mean that such names may be regarded as unrestricted in respect of trademark and brand protection legislation and could thus be used by anyone.

Cover image: www.ingimage.com

Publisher: Südwestdeutscher Verlag für Hochschulschriften GmbH & Co. KG
Heinrich-Böcking-Str. 6-8, 66121 Saarbrücken, Germany
Phone +49 681 37 20 271-1, Fax +49 681 37 20 271-0
Email: info@svh-verlag.de

Printed in the U.S.A.
Printed in the U.K. by (see last page)
ISBN: 978-3-8381-0823-0

Copyright © 2012 by the author and Südwestdeutscher Verlag für Hochschulschriften GmbH & Co. KG and licensors
All rights reserved. Saarbrücken 2012

Contents

Introduction   5

1   Theoretical Background   11

    1.1   Ferromagnetic Order   11

        1.1.1   Exchange Interaction   11

        1.1.2   Band Ferromagnets   12

        1.1.3   Double Exchange   13

    1.2   Micromagnetic Description   14

        1.2.1   Exchange Energy   14

        1.2.2   Zeeman Energy   15

        1.2.3   Stray Field Energy   15

        1.2.4   Anisotropy Energy   16

        1.2.5   Brown's Equation of Static Equilibrium   17

        1.2.6   Micromagnetic Simulations   18

        1.2.7   Static Magnetization Configurations and Domain Walls . .   19

    1.3   Magnetization Dynamics   20

        1.3.1   Field-Induced Dynamics   20

        1.3.2   Spin-Transfer Torque   21

2 Experimental Techniques — 23
- 2.1 Synchrotron Radiation and X-ray Magnetic Circular Dichroism — 24
  - 2.1.1 Photoemission Electron Microscopy — 25
  - 2.1.2 Pump-Probe Technique — 26
- 2.2 Atomic and Magnetic Force Microscopy — 28
- 2.3 Transport Measurements — 29

3 Sample Fabrication and Characterization — 31
- 3.1 Electron and Ion Beam Lithography — 31
  - 3.1.1 Leica Lion LV-1 — 32
  - 3.1.2 Vistec-EBPG5000Plus — 35
  - 3.1.3 Focused Ion Beam Lithography — 36
  - 3.1.4 Alignment System — 38
- 3.2 Electron Beam Resists and Pattern Transfer — 40
  - 3.2.1 Pattern Transfer with Positive Resists — 41
  - 3.2.2 Pattern Transfer with Negative Resists — 43
  - 3.2.3 Pattern Transfer with Shadow Lithography — 44
  - 3.2.4 Patterning on Insulating Substrates — 46
  - 3.2.5 Patterning of Epitaxial $La_{0.7}Sr_{0.3}MnO_3$ (LSMO) — 47
  - 3.2.6 Patterning on Si_3N_4 Membranes — 49
- 3.3 Deposition of Materials — 51
- 3.4 Sample Characterization — 53
  - 3.4.1 Optical Microscopy — 53
  - 3.4.2 Scanning Electron Microscopy — 54

4 Field Induced Domain Wall Motion — 55
- 4.1 Experiment — 56
  - 4.1.1 Magnetic System — 57
  - 4.1.2 Samples — 58
  - 4.1.3 Experimental Technique — 59
- 4.2 Results and Discussion — 61
- 4.3 Conclusion — 69

5 Current-Induced Domain Wall Motion — 71

- 5.1 Displacement in Materials with low Depinning-Field 73
  - 5.1.1 Experiment 73
  - 5.1.2 Results 74
  - 5.1.3 Discussion 78
- 5.2 Time-Resolved Domain Wall Motion 78
  - 5.2.1 Experiment 79
  - 5.2.2 Results 82
  - 5.2.3 Discussion 84
- 5.3 Conclusion 84

6 Spin Configuration in Patterned LSMO — 87

- 6.1 Sample Description and Characterization 88
- 6.2 Magnetic Imaging of LSMO with PEEM 92
- 6.3 Domain Configurations 93
  - 6.3.1 Domains in Square and Triangular Elements 94
  - 6.3.2 Domain Walls in Wires and Rings 102
- 6.4 Field-Induced Nucleation and Depinning 105
  - 6.4.1 Transformation of Domain Walls and Domain States . . . 106
  - 6.4.2 Field-Induced Vortex Core Displacement 110
  - 6.4.3 Field-Induced Domain Wall Depinning and Displacement . 112
- 6.5 Thermally Activated Effects in LSMO 114
  - 6.5.1 Thermal Depinning and Freezing of Domains 115
  - 6.5.2 Thermally Activated Transformation of Domain Walls . . . 120
- 6.6 Conclusion 121

7 Spin Configuration in Patterned Heusler Alloys — 125
- 7.1 Experiment — 126
- 7.2 Spin Configuration in Basic Shapes — 128
  - 7.2.1 Spin Structure in Squares, Rectangles and Disks — 128
  - 7.2.2 Domain Walls in Rings and Nanowires — 129
- 7.3 Thermal Effects — 130
  - 7.3.1 Experiment — 131
  - 7.3.2 Results and Discussion — 131
- 7.4 Conclusion — 133

8 Conclusion and Outlook — 137

Bibliography — 141

Publication List — 147

Introduction

Magnetic materials play the major role in modern mass data storage devices. Commonly used hard drives with spinning disks are cheap, small, and have storage densities that allow for terabyte-sized hard drives in portable computers. Historically, the increase in performance and in storage density has grown exponentially, a trend known as Moore's law [1], and is still the benchmark for recent technology development. The expansion of storage density can be achieved by decreasing the size of the single storage bits and by optimizing the read and writing head for mechanical hard drives. Some milestones of the development of data storage devices are the discovery of the Giant magnetoresistance (GMR) by the groups of Fert and Grünberg [2, 3] and the change from in-plane to out-of-plane magnetized materials. These technological processes allow for the manipulation of smaller domains, increasing the storage density. The drawbacks of this miniaturization are the physical limitations that will end this progress when boundaries are reached. The areal storage density in magnetic hard drives depends, for instance, on the grain size, the bit length and the track width [4, 5]. The thermal stability depends on the grain size and its magnetic volume that decreases with the miniaturization process. Thermal fluctuations can cause a magnetization reversal that leads to the loss of information [4]. For commercial devices, the lifetime should be at least in the range of ten years.

Another way to continue Moore's Law is to replace the recent technology by another one when further improvement is not possible. Recently, solid state drives, based on semi-conductor technology, are entering the hard drive market. In contrast to the common magnetic storage devices, they are free of mechanically rotating or moving parts that can cause failures and decrease the device speed performance. These solid state devices compete with fast writing, reading and access times, and smaller size, but have shorter lifetimes than magnetic devices, and are more expensive (as of 2011). One goal of the development of magnetic structures and materials is to develop a competitive solid state drive based on magnetic materials, the MRAM (Magnetic Random Access Memory). This would combine the advantages of a common hard drive with the benefits of a solid state drive. Different approaches have been suggested, either based on the manipulation by magnetic fields [6] or spin currents [7]. In order to realize domain wall based storage media (e.g. a racetrack or a shift register memory), the fundamental physical processes of the domain wall dynamics in nanometer-sized magnetic wires need to be understood and controlled. Domain walls can be interpreted as quasi particles. Like particles, they exhibit inertia-like behavior when a force is applied, which is particularly important for the understanding of the dynamic displacement of domain walls. During the displacement of a domain wall, deformations of the spin structure occur. This deformation has an influence on the dynamics and can lead to domain wall transformations that can slow down or stop the wall displacement. Local and random pinning as well as edge defects also have a big influence on the dynamics, which in turn affect the depinning field and the critical current density for current-induced domain wall motion.

For a general understanding of the domain wall dynamics, field-induced imaging experiments were performed that reveal the inertial behavior of a domain wall at high temporal and spatial resolution. The domain wall inertia can be seen as a delayed response of the wall motion and in an oscillatory behavior, when relaxing to a ground state inside a potential minimum [8]. Current-induced domain wall motion experiments in smooth nanowires are performed to monitor the influence of edge roughness on the critical current density. A large reduction of the edge roughness and therefore the critical current density was achieved by patterning the nanowires with a negative resist and subsequent ion milling. A linear dependence of the depinning field and the critical current density was seen [9].

Materials with a high spin polarization at the Fermi level are good candidates for spin injectors that can be used for spintronic applications [10]. Also the spin-torque effect is predicted to be more effective in highly spin polarized materials, since the responsible torque scales with the spin polarization. Materials with 100% spin polarization are called half-metallic ferromagnets, where electrons of one spin polarization exists at the Fermi level. Examples for such materials are some of the Heusler alloys, CrO and several of the doped lanthanum manganites.

To utilize these materials, the magnetic properties need to be understood. The crystalline growth, patterning methods, and the material composition are important factors that influence the magnetic behavior of patterned materials and need to be controlled reliably. For a general understanding, magnetic imaging was used to map the spin configuration of patterned half-metallic elements. The influence of magnetic fields and temperature on the spin structure was investigated *in situ*. Reproducible and shape anisotropy dominated structures were found, a prerequisite for domain wall based experiments and applications.

This thesis gives new insights to the field of current- and field-induced domain wall dynamics phenomena in nanopatterned elements and of the role of the different energy terms to the spin configuration of half-metallic systems. These results help to understand and to improve the performance of devices that rely on the displacement of domain walls. Time resolved imaging allows for the determination of the effective domain wall mass and the underlying physical origins. In current-induced domain wall displacement experiments, the influence of edge roughness-induced pinning was identified as having a big influence on the critical current density that is needed to displace a domain wall in a magnetic micro- or nanowire. By applying a specially developed patterning method involving a negative electron beam lithography resist and ion milling, the domain wall pinning at edge defects was highly reduced and the device performance was improved by a factor of four, compared to conventionally produced wires. Patterning methods for the ferromagnetic half-metals $La_{0.7}Sr_{0.3}MnO_3$ (LSMO) and $Co_2FeAl_{0.4}Si_{0.6}$ (a Heusler alloy) were also developed and the spin structure of structured elements are characterized. The epitaxially grown films have a magnetocrystalline anisotropy that influences the spin configuration. The orientation of the patterned elements with respect to the magnetic anisotropy direction was varied in order to study the influence of the anisotropy on the spin configuration in confined elements. In addition to the study of domain configuration in equilibrium half-metallic ferromagnets, the response of the spin structure to applied magnetic fields and temperature was studied. Here, the transformation between different spin configurations (i.e. different energy states) has been shown. This includes as well different domain wall types, as also different domain states. The detailed knowledge about the spin configuration is a prerequisite for further improvements and to design new experiments based on these materials.

The thesis is organized as follows.

Chapter 1 is a short introduction to explain the theoretical background of the origin of magnetism and micromagnetic spin dynamics, which is the fundamental basis for the performed experiments.

Chapter 2 gives an overview of the experimental techniques used in this work. The main focus is on magnetic X-ray imaging, which is used to observe the magnetic spin structure of the patterned magnetic elements.

Chapter 3 deals with the sample preparation, which is key for successful experiments. In addition to the standard patterning techniques (e.g. lift-off), specialized patterning techniques (e.g. involving ion milling or three dimensional structures) that are needed for the experiments are discussed in this chapter.

Chapter 4 gives insight into the first nanoseconds of a field-induced domain wall displacement. Here, a pump-probe technique is employed to image the response of a domain wall with respect to an applied field pulse, with a very high spatial and temporal resolution. The observed delayed domain wall displacement and the oscillatory behavior, when relaxing to the ground state, allow for a determination of an effective domain wall mass. The inertia is caused by the energy transfer between the Zeeman and exchange energy reservoirs.

Chapter 5 reports the results of an investigation of current-induced domain wall dynamics in wires that have very low density of edge defects and therefore less pinning than domain walls in conventionally patterned wires. The low defect density decreases the critical current density by about a factor of four. Time resolved imaging of current-induced domain wall displacement in ordinary lift-off patterned nanowires confirm previously statically imaged measurements.

Chapter 6 details the manifold spin configurations in patterned half-metallic $La_{0.7}Sr_{0.3}MnO_3$ (LSMO). The influence on the spin structure of a uniaxial anisotropy is investigated. Magnetic fields are used to influence the spin configuration *in situ*. Experiments, performed with a heating holder, show thermally activated depinning and the behavior of the spin structure close to the Curie temperature. These studies demonstrate the robustness of the flux closure states to thermal agitation. Thermal depinning and vortex core nucleation is shown in domain walls present in ring elements.

Chapter 7 presents results on patterned Heusler $Co_2FeAl_{0.4}Si_{0.6}$ alloys. In order to control the spin structure in confined elements, the static domain configuration of several element geometries are imaged. The response of the spin configuration to heat is studied, including and thermal vortex core nucleation.

Chapter 8 summarizes the main conclusions of this thesis. In addition, a short outlook is given to put the results in a larger perspective and to discuss possible future experiments and applications.

CHAPTER 1

Theoretical Background

The history of magnetism begins with the use of the mineral magnetite Fe_3O_4, which is called "lodestone" and was most likely magnetized by the earth magnetic field during lava cooling processes [11]. These natural magnets were used over centuries for navigation, while a scientific description was lacking until the early 19th century, when e.g. Ørsted and Biot observed magnetic effects that are induced by electric currents [11]. Nowadays, magnetism, and ferromagnetic materials are still used for navigation but the main applications are in sensors and mass storage devices.

1.1 Ferromagnetic Order

1.1.1 Exchange Interaction

Localized magnetism arises as a consequence of the Pauli principle that requires that the wave function of an electron system is antisymmetric, leading to a coupling between the orbital and the spin components of the wave function. This coupling can be interpreted in terms of the Heisenberg exchange coupling model,

$$H_{ex} = -2J\, \mathbf{S}_1 \cdot \mathbf{S}_2, \tag{1.1}$$

where the exchange integral J is the quantum mechanical overlay of the wave functions of both electrons. For positive J, a parallel alignment of the spins is favored, which leads to ferromagnetism. A negative J causes anti-ferromagnetism, where the spins are aligned antiparallel. To completely describe a many body system with this formalism, the Heisenberg Hamiltonian is given by

$$H_H = - \sum_{ij} J_{ij} \mathbf{S}_i \cdot \mathbf{S}_j. \tag{1.2}$$

Here, every electron combination is taken into account with their respective exchange integral J_{ij} that depends on the distance and also on the relative arrangement in the crystallographic lattice. A relative deflection of the spins to each other from their respective ground state leads to an increased exchange energy. This energy plays a central role in domain walls, where the neighboring spins rotate around a specific angle in a finite area.

1.1.2 Band Ferromagnets

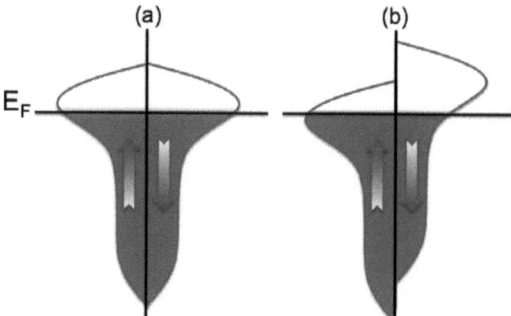

Figure 1.1: Sketches of idealistic electronic bands for (a) a nonmagnetic material and (b) a ferromagnet, where the band-structure has shifted in order to minimize the potential energy.

Magnetism in metals with delocalized electrons cannot be explained by separated electrons with each having a magnetic moment of μ_B. Here, the localized d-electrons can define a magnetic ground state with the non-localized s- and p-conduction electrons as the transmitter of the exchange interaction. One can split the electron band of a metal into the two spin directions (up and down). In Fig. 1.1 the spin-up electrons are in the left band and the spin-down electrons are in the right band. In the case of a non-magnetic metal, both bands are equal and therefore the number of occupied states is equal, as shown in Fig. 1.1 (a).

In a system with a shifted band-structure (Fig. 1.1 (b)) an unbalanced number of electrons in the bands causes a net magnetization **M**.

In the model of band ferromagnets, one can introduce the Stoner parameter [12]. Here the difference of the kinetic and the potential energy of the electronic system close to the Fermi-level is considered. If the energy difference is negative, a ferromagnetic ground state can be achieved. With the Stoner parameter U and the density of states $g(E_F)$ at the Fermi level, the Stoner criteria for ferromagnetism is fulfilled for $U \cdot g(E_F) > 1$. Up to now, all known elements and/or alloys that are ferromagnetic fulfill this criteria. In the case of pure elements, there are only a few that show ferromagnetism at room temperature, such as Fe, Co, and Ni.

1.1.3 Double Exchange

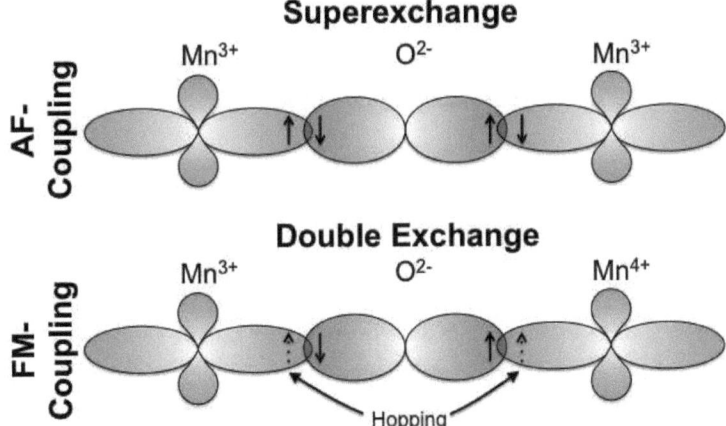

Figure 1.2: *Top*: Sketch of the superexchange interaction. Here, the localized electrons of the Mn^{3+} d-orbital couple with the ones of the p-orbital of the oxygen, which leads to an antiferromagnetic coupling. *Bottom*: The double exchange is illustrated, where the different valency of the Mn ions allow for a hole hopping via the oxygen atom. The first Hund's rule and the conservation of the electron spin during the hopping process favour ferromagnetic order.

In contrast to simple exchange interaction (Section 1.1.1) and band ferromagnetism (Section 1.1.2), ferromagnetic coupling is also possible due to the exchange over a diamagnetic atom. The superexchange in equal valency Mn in MnO couples the Mn spins antiferromagnetically through the oxygen p-orbitals (see top of Fig. 1.2). For different valency atoms, a different interaction called double exchange is visible, which arises from charge transfer between the electrons mediated by the oxygen orbitals, ferromagnetic ordering [13]. Examples of systems that show double exchange coupling include magnetite (Fe_3O_4), and the manganites e.g. LSMO, $La_{1-x}Sr_xMnO_3$ at specific x and temperature ($0.16 < x < 0.5$) [14]. The perovskite structure of these manganites can be described by stacked MnO_6 octahedra that share the corner atoms [11]. The needed difference in the valency can be tuned by replacing the rare earth atom lanthanum with alkali earth, such as Strontium. This replacement of La atoms allows for "hole-doping". The generated holes in the manganese sites lead to a composition of 1-x Mn^{3+} ($3d^4$) and x Mn^{4+} ($3d^3$) [11].

The double exchange involves a charge transport through the occupied oxygen p-orbitals. A hopping mechanism, where an electron of the oxygen goes to one Mn ion, accompanied by an electron transfer from the other Mn ion was proposed by Zener [13]. This hopping over the oxygen atom leads to delocalized electrons in the metal-oxide-metal group. From Hund's rules and the non-allowance of a spin-flip during the hopping process, a ferromagnetic order ensues (see bottom of Fig. 1.2).

1.2 Micromagnetic Description

In large systems, the large number of freedom degrees in the order of number of atoms in the system make an analytical description of magnetism at the atomic scale difficult. To describe magnetism at a semi-classical level, a micromagnetic model was developed by Brown, where a continuous magnetization $\mathbf{M(r)}$ is used to replace the single magnetic moments $\mathbf{S}_i$ [15]. The main energy contribution that influences the magnetization will be described in the paragraph below. The Landau free energy is the sum of these contributions.

1.2.1 Exchange Energy

The exchange energy of neighboring spins is described by Equation 1.1. With the angle θ between the two spins $\mathbf{S}_i$ and $\mathbf{S}_j$, the scalar product of Equation 1.1 can

1.2. Micromagnetic Description

be rewritten for small angles θ as:

$$\mathbf{S}_i \mathbf{S}_j \approx \mathbf{S}^2 - \frac{1}{2}\mathbf{S}^2\theta^2 = \mathbf{S}^2 - \frac{1}{2}(\nabla \mathbf{S} dx)^2. \tag{1.3}$$

For numerical simulations or for larger general problems, the single spins in a volume V are replaced by the M, mean value of the local magnetization. The total exchange energy of the system can now be written as:

$$E_{ex} = \frac{A}{M_s^2} \int dV (\nabla \mathbf{M}(r))^2. \tag{1.4}$$

The exchange stiffness A is proportional to the exchange integral J and M_s is the saturation magnetization. Both parameters are material constants.

1.2.2 Zeeman Energy

Similar to the alignment of a compass needle in the earth's magnetic field, an applied magnetic field forces the spins to be aligned towards the field direction in order to minimize the Zeeman energy. This energy can be written as [16]:

$$E_Z = -\mu_0 \int dV \mathbf{M}(\mathbf{r}) \cdot \mathbf{H}(\mathbf{r}) \tag{1.5}$$

The Zeeman energy can be produced by an external magnetic field induced by coils or permanent magnets. This gives an opportunity to influence the spin configuration and is exploited in many experiments in this thesis.

1.2.3 Stray Field Energy

Starting from the Maxwell equation and the solenoidal vector field $\mathbf{B}$:

$$\nabla \mathbf{B} = 0, \tag{1.6}$$

a magnetic system must compensate any generated stray field with the sum over all dipolar fields. With the stray field $\mathbf{H}_s$ and the magnetic moment $\mathbf{M}$, equation 1.6 can be rewritten as:

$$\mathbf{B} = \mu_0(\mathbf{H}_s + \mathbf{M}), \tag{1.7}$$

which leads to:

$$\nabla \mathbf{H}_s = -\nabla \mathbf{M}. \tag{1.8}$$

This can be written in a Poisson equation with U the scalar potential and ρ the magnetic charge density:

$$\Delta U(\mathbf{r}) = -\rho(\mathbf{r}), \text{ with } \mathbf{H}_s = -\nabla U \text{ and } \rho(r) = -\nabla \mathbf{M}. \tag{1.9}$$

The stray field $\mathbf{M}$ and the potential U can now be determined by solving the Poisson equation. The stray field applies a magnetic field on the magnetization. The resulting stray field energy can be written as [16]:

$$E_S = \frac{\mu_0}{2} \int dV \mathbf{H}_s(\mathbf{r})^2. \tag{1.10}$$

The integral with dV is not limited to the sample size and sums up the stray field over the total space. In order to minimize the stray field energy, the system tries to limit this contribution to the sample dimension. This makes this energy very sensitive to the sample shape and can be used to predefine spin configurations in confined systems.

1.2.4 Anisotropy Energy

While the spins have no direct interaction with the crystal field, the orbital trajectory of electrons around the nuclei induces a relativistic magnetic field that acts on the electron spin, giving rise to the spin-orbit coupling. This in turn gives rise to a coupling between the spin and the crystal lattice and to a new magnetic energy term called the magnetocrystalline anisotropy.

Hence, the orientation of a spin in relation to the three crystallographic axes has a contribution to the total energy that depend on the direction cosine α_i between the magnetization of each spin and the crystallographic axes. This energy density can be approximated by a power series expansion of α_i, with odd powers ruled out (due to symmetry of the energy: $E(\mathbf{M}) = E(-\mathbf{M})$). For uniaxial anisotropy only one direction cosine influences the anisotropy energy density. A power series expansion to the 4th order with the direction cosine α_1 to the z-axis can be written as [16]:

$$E_{ani}^{uniaxial} = K_0 + K_1 \alpha_1^2 + K_2 \alpha_1^4. \tag{1.11}$$

Higher orders of the power series expansion are neglected for further discussion. Negative values for K_1 result in an energy minimum for the uniaxial anisotropy along the z-axis, which now defines an easy direction that favours the spins to be aligned along this axis. Positive values for K_1 create a hard axis in z-direction. The spins now tend to be aligned in the x-y-plane in order to minimize the anisotropy energy. In crystalline cubic systems, a fourfold symmetry of the anisotropy energy density is observed. The power series expansion for this cubic anisotropy up to the 6th order reads [16]:

$$E_{ani}^{cubic} = K_0 + K_1(\alpha_1^2\alpha_2^2 + \alpha_2^2\alpha_3^2 + \alpha_3^2\alpha_1^2) + K_2\alpha_1^2\alpha_2^2\alpha_3^2. \quad (1.12)$$

For positive values of K_1 and neglecting higher orders, a spin alignment in the direction of the three main axis is favored. For negative K_1, the energy minima lie along $\langle 111 \rangle$ directions between the three axis. These examples are idealized cases and in real magnetic systems, different anisotropy contributions may sum up to a more complex effective anisotropy. For many cases, one of these contributions dominates the spin orientation and the other ones can be neglected.

1.2.5 Brown's Equation of Static Equilibrium

The sum of the energy contributions leads to the Landau free energy (at constant temperature):

$$E_{tot} = E_{ex} + E_Z + E_{ani} + E_S. \quad (1.13)$$

This energy is minimized by the equilibrium magnetization distribution $\mathbf{M}_{eq}$. This is the case, when the variation of E_{tot} is zero for the given direction cosines α_i:

$$\delta_{\alpha_i} E_{tot}(\mathbf{M}_{eq}) = 0, \ \forall \alpha_i. \quad (1.14)$$

The equilibrium magnetization that leads to a minimization of the total energy can be determined by the two Brown equations:

$$\mathbf{M} \times \mathbf{H}_{eff} = 0, \quad (1.15)$$

$$\mathbf{M} \times \partial_n \mathbf{M} = 0, \quad (1.16)$$

derived by Brown [15]. Here, $\mathbf{H}_{eff}$ is the effective field with [16]:

$$\mathbf{H}_{eff} = -\frac{\partial}{\partial \mathbf{M}} = \frac{2A}{M_s^2}\Delta \mathbf{M} + \mathbf{H} + \mathbf{H}_s - \frac{1}{\mu_0}\frac{\partial E_{ani}}{\partial \mathbf{M}}. \quad (1.17)$$

The first Brown equation (Equation 1.15) implies parallel alignment of the spins to $\mathbf{H}_{eff}$. The second Brown equation (Equation 1.16) describes the energy at the surface of the magnetic element with ∂_n, the derivative in surface direction. These equations can be used to analytically find the equilibrium spin configuration for simple problems. More complex problems can be solved numerically, where the effective field is calculated for each simulation until $\mathbf{M} \times \mathbf{H}_{eff}$ reaches a predefined threshold value, which needs to be chosen small enough to define a quasi-equilibrium state.

1.2.6 Micromagnetic Simulations

Numerical simulations on micrometer sized magnetic elements are performed to compare experimental findings with theory and to determine the origin of the observed spin structure. The used tool is the open-source program OOMMF, which was developed at NIST [17]. Here a finite difference method is used to solve the Landau free energy and the effective field repeatedly to minimize the energy, until the convergence criteria $\mathbf{M} \times \mathbf{H}_{eff} < \epsilon$ is fulfilled for ϵ sufficiently close to zero [17]. The simulations, performed for this thesis, were ran with the 2D solver of OOMMF, where the magnetization in the z-direction is homogenous, which is an assumption that is only applicable for thin films. In the simulation, the surface of the magnetic element is subdivided in equally sized squares, where every square represents a homogeneously magnetized cell. The size of this smallest element should be chosen in a regime, which is smaller than the exchange length that is typically in the order of 5 nm. Simulations were performed to compare with the findings in Permalloy and $La_{0.7}Sr_{0.3}MnO_3$ (LSMO) thin films. The parameters used are given in the table below.

Parameter	Permalloy	LSMO
Saturation magnetization M_s	800×10^3 A/m	$100 - 400 \times 10^3$ A/m
Anisotropy energy K_1	50 J/m^3	0.01-0.5×10^3 J/m^3
Exchange stiffness A	13×10^{-12} J/m	2.7-3×10^{-12} J/m
Damping constant α	0.02	0.01-0.05

The values for Permalloy are commonly used in the community and are verified by various experiments. The values for LSMO are determined by comparing experimental findings with simulations. Due to the lack of dynamic measurements in LSMO, α is taken to lie between 0.01-0.05.

1.2.7 Static Magnetization Configurations and Domain Walls

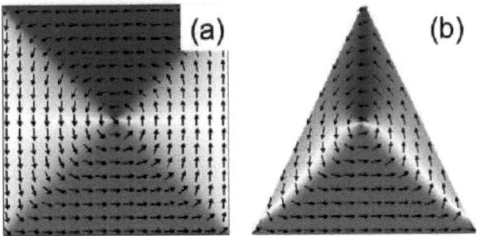

Figure 1.3: OOMMF-Simulation of a Landau-pattern in (a) a square and (b) in a triangle.

Small ferromagnetic particles in the range of a few to some tens of nanometers tend to be in a mono-domain state. This is because the exchange energy at this scale dominates over the dipolar energy associated with the dipolar stray-field at two ends of the element. When the element size overcomes a certain threshold, it becomes energetically more favorable to generate a multi-domain state, like the flux-closure Landau-pattern shown in Fig. 1.3. This flux-closure state can occur in elements with different shapes, for example in a square element as in Fig. 1.3 (a) or in a triangular element as shown in Fig. 1.3 (b).

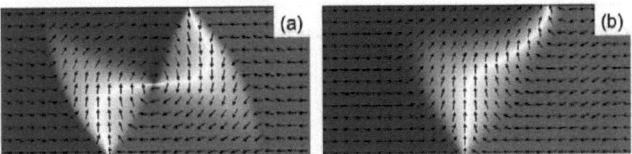

Figure 1.4: OOMMF-Simulation of (a) a vortex wall (VW) and (b) a transverse wall (TW).

Different domains with different spin-orientation are typically separated by domain walls. The spins rotate within these walls, which costs exchange energy. Generally in a one dimensional system, the width of a domain wall is given by [18]

$$\lambda = \sqrt{\frac{A}{K_D}}, \qquad (1.18)$$

where A is the exchange constant and K_D the effective anisotropy.

Two examples of domain walls in a two-dimensional system are presented in Fig. 1.4. The vortex wall in (a) is favored in wide elements, whereas the transverse wall in (b) is predominant in narrow elements.

1.3 Magnetization Dynamics

The magnetization configuration can be modified in different ways. An applied field or a spin polarized current for instance interact with the local spins, thus changing their direction and consequently the spin configuration. A theoretical description of both these ways to influence the spin structure are given in the following.

1.3.1 Field-Induced Dynamics

The response of a magnetic moment to an applied field **H** can be described by [11]:

$$\frac{d\mathbf{M}}{dt} = -\gamma(\mathbf{M} \times \mathbf{H}), \qquad (1.19)$$

where γ is the gyromagnetic ratio. An applied field causes a torque that leads to a precession of the magnetic moment around **H** with the Larmor frequency $\omega = -\gamma \mathbf{H}$ (see Fig. 1.5). Note that the frequency is independent of the perpendicular component $m_\perp$ between **M** and **H**. The angular momentum is preserved by a torque with opposite sign that is applied on the field source by **M**.

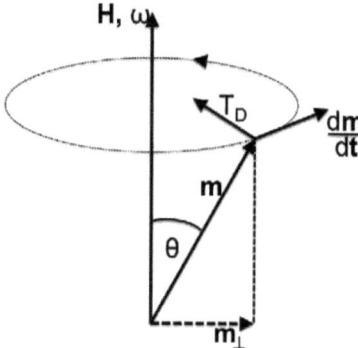

Figure 1.5: The magnetic field **H** generates a torque that forces the magnetic moment to process around **H** with the precessional frequency ω. The additional torque $\mathbf{T}_D$ acts as a damping torque, which leads to a dissipation of the precessional energy.

1.3. Magnetization Dynamics

Equation 1.19 only describes the torque on a magnetic moment, which alone would lead to an infinite precession of the moment around **H**. Experimentally, it is observed that the precession undergoes a damped oscillation and, analogous to a classical compass needle, points in the direction of **H** after the precessional energy is dissipated due to friction. The damping torque $\mathbf{T}_D$ that points towards the axis of **H** (see Fig. 1.5) can be described by [11]:

$$\mathbf{T}_D = \alpha \left(\mathbf{M} \times \frac{d\mathbf{M}}{dt} \right). \tag{1.20}$$

α is a material constant that describes the "friction" and therefore the strength of the damping torque. Combining equations 1.19 and 1.20 leads to the Landau-Lifshitz-Gilbert (LLG) equation of motion. It describes the magnetization dynamics under an applied field **H** [19]:

$$\frac{d\mathbf{M}}{dt} = \underbrace{-\gamma(\mathbf{M} \times \mathbf{H})}_{\text{Precession}} + \underbrace{\alpha \left(\mathbf{M} \times \frac{d\mathbf{M}}{dt} \right)}_{\text{Damping}}. \tag{1.21}$$

This equation has two terms: The precession term that describes the rotation around **H** and the damping term with the phenomenological damping constant α that accounts for the dissipation of energy. The trajectory of **M** can be described as a damped spiral precession around **H** that terminates with $\mathbf{M} \parallel \mathbf{H}$. In magnetic systems, and therefore also for micromagnetic simulations, different fields are combined to the Brown's effective field $\mathbf{H}_{eff}$ that is now used to describe a magnetic system, rather than only a single magnetic spin.

1.3.2 Spin-Transfer Torque

Consider an electron with spin $\mathbf{S}_1$ injected into a ferromagnetic material, which interacts with another electron with the spin $\mathbf{S}_2$. In a closed system where *actio=reactio* is valid for the angular momentum, the conservation of this momentum leads to [11]:

$$\frac{d\mathbf{S}_1}{dt} = -\frac{d\mathbf{S}_2}{dt}. \tag{1.22}$$

This direct interaction between the angular momentum of the two spins gives one the ability to influence the localized spins e.g. with a spin current. A spin current can be generated in any conductive magnetic material, since the conduction electrons will be polarized due to the unbalanced spin states at the Fermi energy.

To describe the influence of a charge current I to a local spin $\mathbf{S}_{1,2}$, Slonczewski added an additional torque term to the LLG equation (equation 1.21) [20]:

$$\frac{\partial \mathbf{S}_{1,2}}{\partial t} = \frac{g}{e} I \mathbf{S}_{1,2} \times (\mathbf{S}_1 \times \mathbf{S}_2), \qquad (1.23)$$

with the electron charge e and the dimensionless prefactor g that depends on the spin polarization. With this additional torque on the magnetization, the Landau-Lifshitz-Gilbert equation of motion reads [21]:

$$\frac{d\mathbf{M}}{dt} = -\gamma(\mathbf{M} \times \mathbf{H}) + \alpha \left(\mathbf{M} \times \frac{d\mathbf{M}}{dt} \right) - \underbrace{(\mathbf{u} \cdot \nabla)\mathbf{M}}_{\text{Extension}}, \qquad (1.24)$$

with $\mathbf{u}$ the generalized velocity

$$\mathbf{u} = \frac{gP\mu_B}{2eM_s}\mathbf{j}. \qquad (1.25)$$

Here, $\mathbf{j}$ is the current density and P the spin polarization. By applying this formula on a magnetic domain wall, Thiaville et al. [21] found that after a critical current density is reached, the domain wall is displaced in the current direction.

So far, an adiabatic torque was assumed. A possible mistracking of the electrons requires the introduction of a non-adiabatic term [22]. The extended LLG equation with the adiabatic and the non-adiabatic term is now suited to describe the magnetization dynamics for applied fields and spin currents and reads:

$$\frac{\partial \mathbf{M}}{\partial t} = -\gamma(\mathbf{M} \times \mathbf{H}_{eff}) + \alpha \left(\mathbf{M} \times \frac{\partial \mathbf{M}}{\partial t} \right) - \underbrace{(\mathbf{u} \cdot \nabla)\mathbf{M}}_{\text{Adiabatic}} + \underbrace{\beta \mathbf{M} \times ((\mathbf{u} \cdot \nabla)\mathbf{M})}_{\text{Non-Adiabatic}}. \qquad (1.26)$$

Experimentally, current-induced domain wall motion is observed in various experiments, but detailed knowledge for example of the non-adiabaticity factor β is still lacking. It is expected that the spin torque is more efficient for large spin polarizations P and small saturation magnetization M_s [23]. This makes the class of ferromagnetic half-metals very interesting, as they fulfill these criteria, providing a spin polarization of up to 100% and for some material compositions a very low saturation magnetization.

CHAPTER 2

Experimental Techniques

Different experimental techniques for magnetic characterization are introduced in this chapter. The focus is set on techniques that allow a direct measure of the local spin configuration, and in particular on techniques, that are based on magnetic imaging techniques, e.g. magnetic force microscopy (MFM) and X-ray microscopy, which is required for the experiments conducted during this work. These powerful nonintrusive techniques enable fast and reliable imaging of the local magnetization, a prerequisite for many experiments. Magnetoresistance effects are used, for instance, to detect a domain wall inside a nanowire. Techniques that are based on these effects are ideally suited if cryogenic temperatures are required in the experiment.

MFM images, taken with a "Digital Instruments Dimension 3100", map the spin structure in order to characterize samples prior to X-ray microscopy experiments and to optimize material parameters. This stray field sensitive technique is not ideally suited for in-plane anisotropy materials with flux closure spin configurations. However, the availability and the reliability of this machine made it a useful tool.

Magnetic imaging techniques based on X-ray magnetic circular dichroism (XMCD) combine high spatial resolution of X-ray based microscopy and the time resolution that can be achieved by triggering an excitation in the experiment making use of the short light pulses ($\sim$50-70 ps) of the synchrotron light source. With these techniques, both the static and the dynamic magnetization configurations can be mapped. Magnetic imaging using the XMCD effect are the basis for most of the experiments presented and various X-ray microscopy measurements are performed at various synchrotrons.[1]

2.1 Synchrotron Radiation and X-ray Magnetic Circular Dichroism

Some techniques that are used to image magnetic materials exploit X-ray magnetic circular dichroism (XMCD). This effect leads to a difference of X-ray absorption depending on the sample magnetization and the photon polarization. At the L_2 and the L_3 edge of a ferromagnetic metal, e.g. Fe or Co, the absorption due to the imbalance of spin-up and spin-down electrons causes a difference in the absorption of circularly polarized light, which can directly be measured by counting the emitted photo electrons.

When a 2p-electron is excited by circularly polarized light, it can be lifted to the Fermi energy. The ensuing relaxation of a higher core electron generates an Auger electron that has sufficient energy to excite a cascade of secondary electrons that can leave the sample surface and are therefore measurable. The total yield of secondary electrons is proportional to the absorption, and provides a measure of the local spin orientation.

To utilize this effect, one needs high intensity monochromatic, and circularly polarized X-ray light. These special requirements are available at synchrotron light sources. 3rd generation synchrotron light sources have the best conditions to provide all these needs. In these facilities, high intense electron packets (bunches) are accelerated to velocities close to the speed of light. These bunches are kept in a storage ring with a large diameter, e.g., 288 m at the SLS in Switzerland.

---

[1]Experiments during my PhD time were performed at the following synchrotrons:
- BESSY II, Berlin, Germany
- ELLETRA, Trieste, Italy
- SLS, Villigen, Switzerland
- Diamond, Oxford, U.K.
- ALS, Berkeley, U.S.A.

2.1. Synchrotron Radiation and X-ray Magnetic Circular Dichroism

When the trajectory of charged particles is bended, synchrotron light emerges. This broad banded light becomes more energetic and more intense the faster the particles and the smaller the curvatures are. To keep the beam in a quasi circular course, bending magnets are used. These rather simple deflection devices produce light that can be used for XMCD experiments after monochromizing it. These monochromators use grating mirrors at a grazing angle of incidence. With the broad spectrum of emitted light, the bending magnets are sufficient for some experiments, but not the best choice if a high monochromatic photon flux is required.

A more specialized tool is the insertion device, also called undulator. Here, two arrays of permanent magnets are brought close together with the electron beam between them. Depending on the gap and the relative phase of these arrays, the energy and the desired polarization of the generated X-ray beam can be defined. The spectra of the emitted photons has now a well defined peak, that can be selected with a monochromator. In order to detect the spin polarization of a sample, one needs the information of both helicities, circular plus and circular minus. An undulator is motorized and has the ability to move the phase of the permanent magnet array to a position that results in a reversed light helicity.

2.1.1 Photoemission Electron Microscopy

One imaging technique that is used to map the spin-configuration of a sample is XMCD-photoemission electron microscope (XMCD-PEEM). A photograph of the PEEM tool at the SLS (SIM-beamline) is shown in Fig. 2.1. Photoelectrons are generated by the X-ray beam, hitting the sample at an angle of 16°. The secondary electrons are magnified and imaged by electron optics, consisting of different lenses. The objective lens, directly in front of the sample, is set to an electric potential of 15-20 keV above the sample potential (the high voltage, HV, is supplied by the HV connection, Fig. 2.1_1). The high voltage is needed to accelerate the electrons towards this lens, and can lead to discharges between the sample and the objective lens. Such a short but high intense discharge mostly hits the cap of the sample holder, but can also damage or sometimes re-magnetize the magnetic structures. There is no way to absolutely avoid discharges, but low pressure (ideally below $5 \cdot 10^{-9}$ mbar), a clean and smooth sample and sample holder surface, a reduced acceleration voltage, and an increased distance to the objective lens can help to reduce the discharge probability.

Figure 2.1: Photograph of the SLS PEEM at the SIM-beamline. Main components are labeled from 1-8.

The whole sample stage can be rotated to allow for different contrast directions (Fig. 2.1_2). After passing through the objective lens, the optical quality of the beam is enhanced by moving a contrast aperture into an internal focus point (Fig. 2.1_3). After passing by an energy analyzer (Fig. 2.1_4) and an energy slit (Fig. 2.1_5), the electron beam is widened by magnetic lenses and accelerated onto a micro channel plate (Fig. 2.1_6). This plate multiplies the incoming electrons by an avalanche effect inside micrometer sized tubes. At the end of this plate, the electrons hit a scintillator crystal, which generates visible light that is detected by a CCD-camera (Fig. 2.1_7). The amplification can be tuned by a voltage that is applied to the channel plate. This can be exploited to trigger the amplification in time resolved pump-probe experiments that are described in Section 2.1.2. For prelinearly sample alignment, the sample is illuminated by a Hg-lamp rather than by X-rays (Fig. 2.1_8).

Two images taken at opposite helicities can be combined to yield a magnetic contrast parallel to the X-ray beam. Due to the small out-of-plane component of the X-ray beam, PEEM is also slightly sensitive to the out-of-plane magnetization.

2.1.2 Pump-Probe Technique

The synchrotron has many electron bunches (multi-bunch mode). The time distance between two bunches is typically 2 ns, which is equivalent to approximately 60 cm at the speed of light and would allow theoretically for 480 single bunches inside the storage ring of the SLS.

For practical reasons, only 390 of the possible 480 electron bunches at the SLS are filled with electrons and a gap is left empty. In the center of this gap, a single

2.1. Synchrotron Radiation and X-ray Magnetic Circular Dichroism

but high intense electron bunch, the camshaft bunch, is positioned. This is used for time resolved pump-probe measurements, as the experiment can be gated to the isolated X-ray pulse.

Pump-probe techniques are a powerful method to perform time resolved measurements. In particular when one-shot experiments are not possible or the acquisition time is too long, a repetitive experiment gives the option to accumulate data for well defined time windows.

In time resolved PEEM experiments, the time before and after the camshaft bunch is needed, as the time required to trigger the channel plate takes a few nanoseconds. The trigger is synchronized with the orbit clock of the synchrotron beam and adds additional 280 V to the channel plate, which is set to a lower base value. With this setup, the multi-bunches are gated out. The temporal resolution is now only limited by the width of the of the high intensity single bunch, in the range of 50-70 ps.

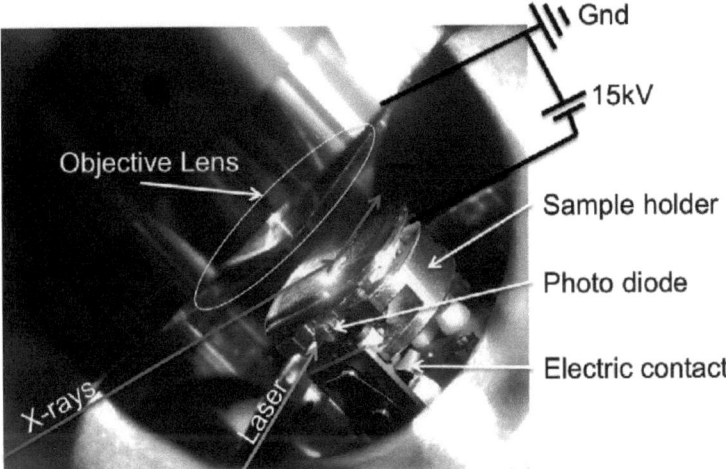

Figure 2.2: Top view of the objective lens and the experimental setup inside the PEEM for field-induced domain wall motion.

By triggering the experiment to the frequency of the synchrotron, only a small time window equal to the width of the single bunch is observed. The observed moment can be varied in time by setting a relative delay between the pulse (pump) and the trigger (probe). High frequency excitation inside the PEEM is problematic, since no high frequency cables are available between the high voltage rack and the vacuum chamber. The method of choice to deal with this is a laser pulse to generate free charge carriers in a photo diode that is mounted directly on the sample holder inside the vacuum. Very short current pulses with very steep rise times can be generated. Thus the typical characteristics of such a pulse depend on the diode used, the laser power and, the sample impedance.

A big draw-back of this technique is the sensitivity to discharges. Damage to the fragile photodiode can be avoided by adding a suppressor diode on the circuit board. Using this arrangement, any incoming discharge will be redirected and flows to ground (the more sensitive magnetic element is usually damaged, since too much current is involved). In very rare cases, an element might survive a discharge, but normally a new sample needs to be mounted. This unfortunately consumes a lot of time, as the clean vacuum conditions must be obeyed.

The experimental setup for field-induced domain wall motion is shown in Fig. 2.2. Here, the distance to the objective lens is large compared to usual PEEM measurements, in order to prevent possible discharges. The image shows the photo diode and one electric contact. The suppressor diode is on the back of the sample holder and therefore cannot be seen.

2.2 Atomic and Magnetic Force Microscopy

Magnetic force microscope (MFM) is a scanning microscope similar to an atomic force microscope (AFM). The difference between these two microscopes is the used of a magnetized tip in MFM. Here, every line is scanned twice. First, the topological surface is scanned and secondly after lifting the tip at a well defined height above the surface, the stray-field is measured.

This technique is suited for magnetic out-of-plane anisotropy materials, but it is also sensitive to the stray field generated by domain walls in in-plane magnetized materials. Prior to a beamtime measurement, or on insulating samples, where for example PEEM measurements are not possible, the knowledge of the static magnetic properties of the sample is very helpful. MFM images of a Heusler alloy on an insulating MgO substrate are shown in Fig. 7.1 and Fig. 7.2 in Chapter 7.

2.3 Transport Measurements

In addition to imaging techniques, transport measurements were performed in order to gain insight into the electronic properties of the sample. Different magnetoresistance (MR) effects can occur in magnetic samples. Some of them are small and therefore the resistivity is typically measured in a four-point geometry. Many experiments were performed in order to determine the position of the domain wall inside the bend of a zig-zag wire. The existence of a domain wall between these contacts will cause a change in resistivity in most magnetic materials.

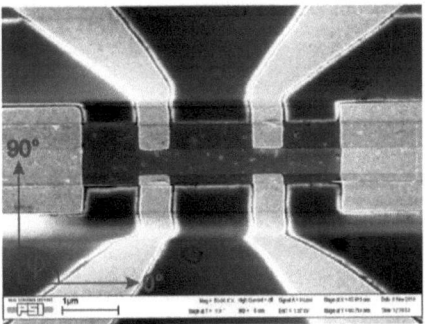

Figure 2.3: SEM image of a patterned Heusler alloy bar, which is connected with four Au-contacts in a Hall geometry. The insulating substrate (MgO) leads to charging effects in electron based microscopy techniques (e.g. SEM and PEEM), here manifest in image distortion and horizontally orientated stripe-like artifacts.

The origin of the decrease in resistivity for domain walls in Py is for example the magnetoresistance (MR) effect, which depends on the alignment of the spins to the direction of the electron flow. A domain wall reduces the number of spins that are aligned parallel to the electron current and therefore reduces the resistivity [24]. Transverse and vortex domain walls have different spin structures, which gives rise to different resistance values. The changes are very small and the absolute domain wall type cannot easily be determined by these measurements. Additionally, domain walls can transform to unpredictable multi domain states with complex spin structures during current pulses [25, 26], which makes a clear determination using the resistance value only very difficult [9].

A typical geometry for transport measurements is shown in Fig. 5.1 (a) in Chapter 5, where the domain wall can be detected due to the AMR effect. Effects such as the planar Hall effect can be measured in the Hall-geometry, shown in Fig. 2.3.

CHAPTER 3

Sample Fabrication and Characterization

Sample preparation is a fundamental step towards successfully performing experiments on magnetic micro- and nanostructures, as suitable samples with the desired properties are needed. Important parameters to consider are the right choice of substrate, design, magnetic material, electric contacts, patterning method and so on.

In this Chapter, the patterning of nano- and micrometer-sized magnetic elements will be described, covering different electron and ion beam lithography systems. In addition to standard methods like pattern transfer with lift-off technique, specialized techniques were developed in the framework of this PhD [8, 9, 25–39].

3.1 Electron and Ion Beam Lithography

The choice method to generate individual patterns with resolutions down to a few tens of nm is electron beam lithography (EBL) or in some cases focused ion beam (FIB) lithography. In Section 3.1.1 and Section 3.1.2, two different EBL systems and in Section 3.1.3 an FIB- lithography system will be introduced. An introduction to the software that is needed to prepare an exposure will be given for two individual EBL systems.

The general working principe of EBL is that an electron sensitive resist is spin-coated on the frontside of a chip and after exposure to an electron beam, the molecular properties of the exposed areas are changed in such a way that either the exposed or the unexposed area can be dissolved and removed in a special developer. Generally one differentiates between positive (see Section 3.2.1) and negative (see Section 3.2.2) resists, according to whether the exposed resist is removed or remains after the development. This is analogous to optical lithography, where the resist is exposed with photons instead of electrons. Optical lithography plays a minor role in my sample preparation and is not discussed further.

With EBL, the desired pattern is generated by scanning the electron beam on the surface to expose the required area, similar to the working principle of a scanning electron microscope (SEM), which in many cases can also be used for lithography. This scanning of the beam can be performed by electronically deflecting the beam in a certain area, which is called a stitching field. If an exposure bigger than this field needs to be exposed, the exposure is usually split in several parts, that are then "stitched" together, or the table is moved during the exposure and the deflection of the beam is corrected accordingly (continuous path control). The precise overlay of the single stitching fields is very important and requires a very precise interferometric alignment of the sample stage and also a good beam alignment, which is one of the main advantages of a specialized EBL system, compared to an ordinary SEM.

3.1.1 Leica Lion LV-1

The Leica Lion LV-1 is a very special EBL system, which has a low-voltage (LV) operating mode. The LV mode (running at 2.5 keV) has the advantage of a negligible proximity effect (where the area next to an exposed area is exposed by secondary electrons), but the disadvantage of long exposure times due to the low electron current. A scheme of the electron optics of the Leica Lion LV-1 is shown in Fig. 3.1. The electrons are generated by a zirconia (ZrO) coated tungsten (W) thermal field emission cathode at a temperature of 1600°C and a pressure below 10^{-8} mbar. The electrons are accelerated and pass the beam blanker and the optical elements after they exit the column at the objective lens. The beam blanker uses a high voltage plate to deflect the beam away from the sample. This is used to switch the beam "on and off" between line segments and between writing the different patterns.

This system is equipped with a continuous path control. This means that long lines can be exposed without stitching errors. The desired line width can be tuned by adjusting the exposure dose and the focus of the beam. This EBL tool was

3.1. Electron and Ion Beam Lithography

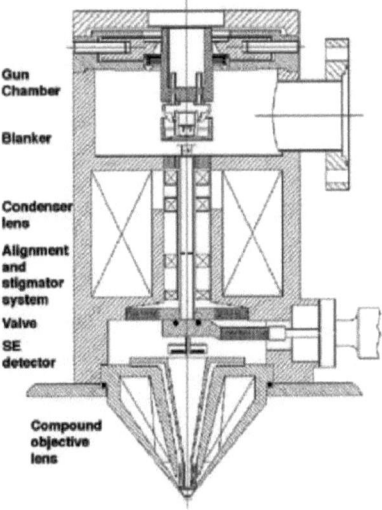

Figure 3.1: Column of the Leica Lion LV-1 electron beam lithography system [40].

used for the first two years of my PhD and to ease the preparation of individual exposure data, the program Bez-Maker was written. The program and the used data format is described next:

Data format and Layout Designer: BEZ-Maker

Before the program "Bez-Maker" was written, the exposure data for the Leica Lion LV-1 was usually prepared by specialized macros that only fulfill a specialized task.

The principal data format is called ".bez" which stands for Bézier.[1] A Bézier curve is a special type of curve that in the simplest case consists of two points that describe a simple line (a linear Bézier curve, shown in Fig. 3.2 (a)). By adding a third base point, a curved line is generated between the two starting points P_1 and P_3, where the derivative in the vicinity of the starting points points into the direction of the base point P_2 (Fig. 3.2 (a), a quadratic Bézier curve). A cubic Bézier curve with four base points P_1-P_4 is shown in Fig. 3.2 (a) on the right (blue curve).

---

[1] After Pierre Bézier, who discovered this mechanism in the 1960's, in parallel to Paul de Casteljau.

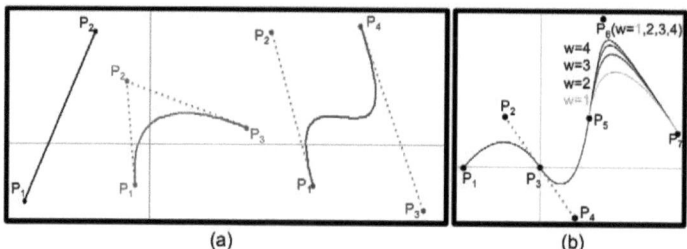

Figure 3.2: (a) Examples for a linear (black), a quadratic (red) and a cubic (blue) Bézier curve. The dashed lines are guides to the eye and indicate the first derivative of the curve in the starting- and the end-point. (b) Example for a spline, defined by three quadratic Bézier curves. The spline is differentiable between the points P_1 - P_7 if the pairs of weighted points (P_2-P_4 and P_4-P_6) are point reflected to the corresponding start/end point in-between each pair (P_3 and P_5, respectively). The weight of P_6 varies from 1-4 (w=1 is printed in turquoise). The smooth curves (w=1-w=4) through P_5 demonstrate the weight independence of the differentiability for Bézier splines.

A Bézier curve of order n can be described by:

$$C(t) = \sum_{i=0}^{n} B_{i,n} P_i \ , \ t \ \epsilon [0,1], \tag{3.1}$$

where $B_{i,n}$ is the Bernstein polynomial:

$$B_{i,n} = \binom{n}{i} t^i (1-t)^{n-i}. \tag{3.2}$$

All points can have a quasi mass which attracts the curve towards this point and therefore deform it. An example of varying weight on a Bézier curve is shown in Fig. 3.2 (b) at point P_6, where the weight w varies between $w = 1$ and $w = 4$. These rational Bézier curves are described by

$$C(t) = \frac{\sum_{i=0}^{n} w_i B_{i,n}(t) P_i}{\sum_{i=0}^{n} w_i B_{i,n}(t)}. \tag{3.3}$$

The Leica Lion LV-1 system only accepts weights equal to **1** for the start and the end point and Bézier curves up to an order of four (two central base and one start- and one end-point). In general Bézier curves can have weights different from 1 for every base point.

A spline is a sequence of Bézier curves and allows a smooth, complex curve to be generated (i.e. the derivative of the curve can be calculated at every point).

3.1. Electron and Ion Beam Lithography

Fig. 3.2 (b) shows a spline, composed of three quadratic Bézier curves. The Leica EBL tool is able to combine lines with the same end- and start-point so that the beam does not need to be blanked inbetween. This is very useful for long and narrow lines, as this prevents stitching errors that may cause a break in the line.

A large diversity of different samples needed to be produced for the group and therefore a specialized program to develop and generate these Bézier-files was written: the Bez-Maker. The user friendly environment of Microsoft Visual Basic was used to generate a vector based program that handles Bézier curves and also allows for direct manipulation of these curves (the presented curves and splines in Fig. 3.2 are generated with this software). Some of the main features are the following:

- Curves with the same dimension and the same start- and end-point are automatically combined (for example a spline).

- The points can be moved individually to deform the curve.

- All curves are visualized so that a modification of the elements is easy.

- In addition to the Bézier-, also ".slo"-Files can be read in. These files contain the whole exposure which is a collection of ".bez"-files, their relative coordinates and beam settings.

- Selected vector based files like the ".ai"-format (Adobe Illustrator) can be converted into the ".bez"-format, so that patterns can be generated with other programs as well.

- The total length of the Bézier curves can be determined, allowing an estimate of exposure time.

Structures, generated by the Leica Lion LV-1 that make use of the advantage of splines and higher order Bézier curves are shown in Figs. 3.5, 3.4, 3.11 (c) and 3.11 (d).

3.1.2 Vistec-EBPG5000Plus

In addition to the Leica system a new dedicated EBL system (Vistec EBPG5000Plus[2]) was used. This system can perform exposures with higher electron energies (up

---

[2]EBPG stands for Electron Beam Pattern Generator

to 100 keV), which allows the exposure of thicker resists. In this tool, beam alignment is performed automatically and the writing time is very fast due to the high available electron current.

The height of a sample in the tool is limited to a fixed position but allows deviations up to 50 μm. It is easier to fix the height of the sample rather than readjust the tool settings for the focus. Therefore, the sample surface must be within this tolerance. To facilitate this, different mounting positions are available. If the required height is not available, thin Si chips can be used underneath the sample to lift it or thin Ti plates can be used at the spring clip of the sample holder in order lower the sample height. Compared with the older Leica EBL tool, the setting up time is considerably reduced. The reliability and advantages of this system resulted in a complete changeover from the Leica system to the EBPG5000Plus.

Layout Design: L-Edit, Layoutbeamer, C-Job
The exposure data for the EBPG was generated by the program L-Edit and was further converted by the program LayoutBEAMER. This program sets the precision/step-size for the exposure and is also able to perform a proximity effect correction. This is very helpful, as the proximity effect on bulk samples cannot be ignored for large elements and especially for the high acceleration voltage of 100 keV.

The last step uses the program C-Job. Here, different files, which are prepared with LayoutBEAMER, can be arranged in one exposure. All exposure parameters are set in C-Job. Alignment positions can also be defined. This is very important for multi-step exposures. The alignment is carried out completely automatically during the exposure with an accuracy down to 30 nm.

3.1.3 Focused Ion Beam Lithography

A different approach to the electron beam lithography is to directly pattern nano-sized magnetic elements, using Focused Ion Beam Lithography (FIB). Here, a focused Ga beam is scanned over the sample surface to remove material, or to modify the surface. Ga ions can reduce or destroy the magnetic properties when implanted in magnetic materials [41]. Therefore, it is possible to define a magnetic structure without removing the material. The system used is a dual-beam system at the University of Konstanz. The sample is rotated by an angle of 54° to face the FIB column. The sample is brought to a position where the electron beam

3.1. Electron and Ion Beam Lithography

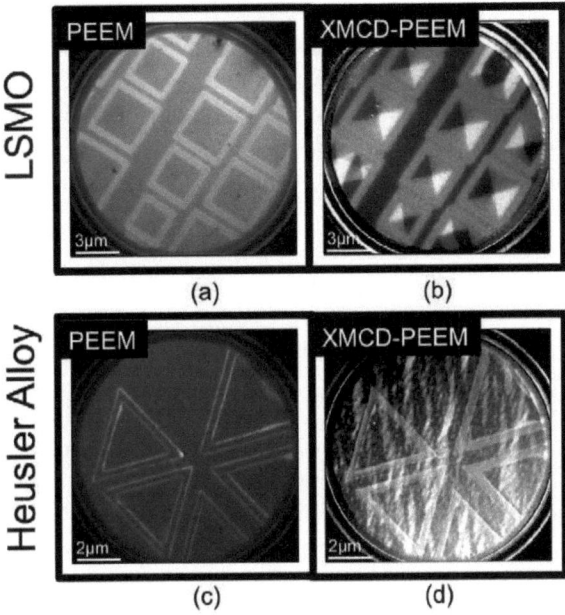

Figure 3.3: (a) FIB structures on a 50 nm thick LSMO film. The patterned lines are nonmagnetic, resulting in a shape anisotropy defined domain structure in (b). (c) FIB structures on a Heusler alloy film. The FIB dose was not sufficient to remove the film or to turn it into the paramagnetic state at the FIBed region, and the stripe domains in (d) cross the patterned lines.

and the ion beam meet at the "eucentric height". With a good alignment of these two beams, the ion beam position can be set using the SEM without exposing the sample to Ga ions. This is particularly useful if the sample is sensitive to ion irradiation and the surface around the patterned area should not be treated. The exposure data is generated in L-Edit end exported to ".gds". The exposure is then performed by a Raith system. By setting the coordinates of the exposure, the desired exposure dose, and the stitching field, the exposure can be started. In Fig. 3.3, two examples of FIB patterned films are shown. In (a) the FIB lines are visible on 50 nm thick LSMO film and in the associated XMCD image (b), where the shape anisotropy defined magnetic domain patterns can be observed, referred to as a Landau pattern. The FIB lines in (c) are written with a very high dose in a Heusler alloy film. The presence of ripple domains in (d) indicates little influence of the Ga beam exposure and that the patterning was not successful.

3.1.4 Alignment System

For many samples, a precise overlay of different exposures is extremely important if especially small structures need to be contacted. For the Leica Lion LV-1 EBL system, crosses are used as alignment marks. This has the advantage that one can very precisely determine the origin position of two or three crosses to finally rotate and shift the coordinate system of the second exposure so that the second step matches the first step. The rotation of any arbitrary angle can be applied by the program Bez-Maker (see Section 3.1.1). A further advantage of the low voltage mode is that the alignment marks can be made of virtually any material as the material contrast at 2.5 keV is very high. Usually, the alignment marks are written in the same lithography step as the magnetic structures and therefore consist of the same material, which saves an additional lithography step and increases the overlay precision. An advantage of the Leica system is that any other arbitrary shape (i.e. any structure on the sample) can be used to align the coordinate system to that of the sample. In Fig. 3.4 (a), the overlay of Py-nano-wires on a single graphene flake is shown, aligned by Au crosses, that were added by optical lithography. An overlay of a non-local spin-valve is shown in Fig. 3.4 (b) and an overlay of Au contacts on Py wires is shown in Fig. 3.5. In Fig. 3.5 (a), the grids above and below the wires are used to define the coordinate system.

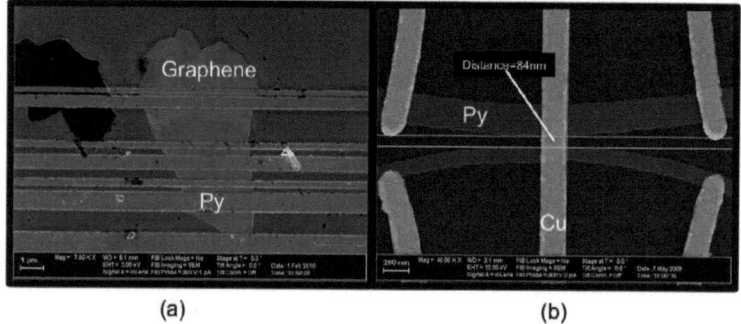

Figure 3.4: (a) Overlay of Py-stripes on a randomly placed graphene flake, aligned on optically defined Au crosses. (b) Close up of a non local spin-valve. The overlay of the Cu-contacts is carried out by manual alignment on Py crosses, which are written and deposited in the same step as the nanowires.

In contrast to this manual alignment, the fully automated alignment procedure of the EBPG can handle a broad range of different pre-defined alignment marker

3.1. Electron and Ion Beam Lithography

shapes. For simplicity, only squares with different sizes where used (typically sizes are squares with an edge length of 10, 15 and 20 μm). The rough coordinate of a pre-alignment marker is determined using an optical microscope with an accuracy of around 20 μm. The EBL system is able to detect this marker and automatically finds further markers to perfectly align the single sub-exposures. The rotation of the sample is corrected automatically in a certain range by the system, which requires a very good prepositioning of the sample on the holder.

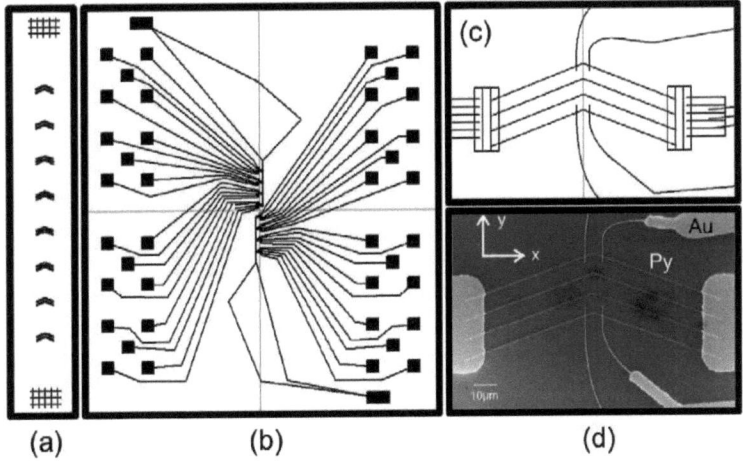

Figure 3.5: (a) Design of the first lithography step for nanowires, written in HSQ on a Permalloy thin film. (b) Overlay design with bond pads and cables. The coordinate system for the alignment is orientated on the crosses in (a), above and below the wires. (c) detailed view of the Bézier design. (d) Finished sample with MR-contacts at the top and bottom Py kink and electric contacts (left and right).

The disadvantage, due to the high acceleration voltage of the EBPG, is that not all alignment markers show a high contrast. Best results are achieved for markers made of 60 nm gold in thickness, which requires an additionally lithography step and consumes gold. To reduce the valuable EBL preparation time and to spare the gold used in the evaporation, complete 4"-wafer were pre-patterned with a standardized alignment system. On these wafers, 18 mm × 18 mm chips were predefined to be diced into single chips after the gold deposition (for thermal deposition see Section 3.3). This introduction of a standardized alignment marks additionally simplified the data preparation, since it was then used for all exposures. In addition to the squares, arbitrary shapes can also be aligned, but the accuracy is limited by the size and contrast of the marker used. A simple SEM mode is implemented at the EBPG system, which can be used for samples without suitable alignment markers, for example when the edge definition is too poor for an automatic alignment or when other alignment markers were provided.

3.2 Electron Beam Resists and Pattern Transfer

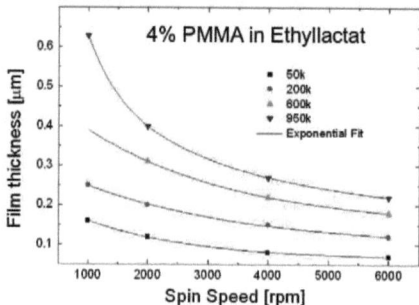

Figure 3.6: Spin curves for different molecular weights of PMMA in Ethyllactat. The red fits are first order exponential decays and a second order exponential decay for the 950 k resist.

In the process of a sample preparation, the pattern in the resist needs to be transferred to the desired material and many different approaches and techniques exists. Electron beam resists are organic or inorganic chemical compounds, dissolved in a solvent, which can be coated on the substrate surface using a spincoater. The photoresist is placed on the sample with a pipette. The subsequent rotation of the sample removes the excess resist leaving a homogeneous film with a well defined thickness remains (along the sample edges, the resist is slightly thicker). The thickness of the resist depends mainly on the spin speed, the spinning time, the solvent used, and the type of the resist but also on the substrate material and its size. A typical spin curve is shown in Fig. 3.6, with exponential fits to interpolate the thickness between the data points.[3] After spin coating, the resist is baked on a hot plate. This is needed to evaporate the solvent and in some cases to change the physical properties of the resist. For example PMMA needs to be heated above the glass transition temperature T_{Glass}(PMMA) of around

---

[3]The data points are taken from Allresist GmbH

3.2. Electron Beam Resists and Pattern Transfer

390 K [42], where the polycrystalline resist transforms into the amorphous phase, resulting in a smooth resist surface.

3.2.1 Pattern Transfer with Positive Resists

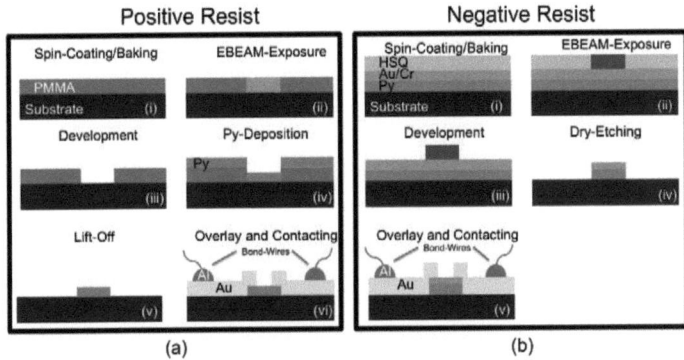

Figure 3.7: (a) Pattern transfer using a positive resist (PMMA). The resist is coated on the substrate (i) and afterwards exposed by electron beam lithography (ii). After developing the exposed resist (iii), the metal is deposited on the sample (iv). The metal on the PMMA is removed during the lift-off (v). A second exposure is used to prepare Au-Contacts for electronically contacting the structure (vi). (b) Scheme of pattern transfer, using a negative resist. (i) The magnetic material and a buffer layer (Au/Cr) is deposited on the substrate prior spin coating the resist (HSQ). The pattern is written by electron beam lithography (ii). After developing, the exposed resist remains (iii). A dry-etching recipe that removes the HSQ and some of the buffer layer transfers the pattern from the HSQ into the magnetic material (iv). An overlay step with a positive resist pattern transfer like in (a) is used to generate the electronic contacts (v).

Positive resists have the characteristic that the exposed area dissolves in a solvent, whereas the unexposed area remains. The most common positive EBL resist that was used is PMMA (Polymethyl methacrylate). The electron sensitivity of this resist depends on the thickness and on the molecular weight. Available resists are typically in the range of 50 k to 950 k, with 50 k being the most sensitive resist. The schematics of patterning a sample with a positive resist is shown in Fig. 3.7 (a). This pattern transfer method is used in most of the samples, where no specialized requirements are needed and for adding electric contacts and bond pads. A typical preparation process for a sample with Au contacted Py elements prepared with PMMA and lift-off technique is given below and shown schematically in Fig. 3.7 (a), steps (i)-(vi):

1. PMMA is spin-coated on the sample. For this, the sample is put on a chuck, which has a hole so that the connected vacuum holds the sample in place during spinning. Before starting the spin coater, the resist is placed on the sample surface with a pipette. The spin speed is taken from a spin curve of this resist (a typical spin curve for PMMA dissolved in ethyllactate is shown in Fig. 3.6). The resist thickness used depends on the particular purpose. To avoid lift-off problems, a rule of thumb is to use a resist that has twice the height of the subsequently deposited thin film. If the sample is sputter etched before the deposition, the resist height may decrease during the sputtering and needs to be adjusted accordingly.

2. Directly after the spin coating, the sample is baked for 1 min on a hotplate at a temperature of 170°C (Fig. 3.3 (a), step (i)).

3. After preparing the exposure data and loading the sample into the EBL system, the individual design is exposed into the resist (Fig. 3.3 (a), step (ii)).

4. The exposed PMMA is developed using an automatic Hamatech development tool (Steag-Hamatech HME 500). Methyl-iso-butyl-ketone (MIBK) and isopropyl alcohol (IPA) in a composition of 1:3 MIBK:IPA (60 s for $\sim$150 nm resist) is used as a developer for high resolution features.
1:1 MIBK:IPA is used when the resolution plays a minor role and thick resists needs to be developed. The Hamatech tool sprays the developer homogeneously on the rotating sample. After this, the sample is cleaned with 30 s IPA and spin-dried at 3 krpm.

5. The success of the exposure for structures is determined optically (see Section 3.4).

6. If a very clean surface is required, residual resist can be removed by ion-etching or with an oxygen plasma, which removes the organic residue on the surface. The substrate at the exposed area is now ready for material deposition (Fig. 3.3 (a), step (iii)).

7. The desired material can now be deposited on the whole surface of the chip (see Section 3.3). The rest of the chip is covered with the unexposed PMMA. If regions need to be deposited with another material, areas that do not require metallization are typically masked with a piece of alumina foil, or blanked with a shutter (Fig. 3.3 (a), step (iv)).

8. After deposition, a lift-off step is necessary to remove both the PMMA and the undesired metal on top of it. The lift-off for bulk samples is carried out by leaving the sample in an acetone bath in ultra-sound. Samples that are fragile (e.g. membranes) or structures that might be damaged or removed by ultrasound (e.g. graphene flakes) are handled with more care, by putting them upside down in an acetone filled beaker with a stirrer [43] (Fig. 3.3 (a), step (v)).

9. To add electric contacts on the structures, the steps 1-8 in this list are repeated with an exposure that contains the necessary wiring and the bond pads. In addition, an overlay procedure is needed in order to align the second exposure exactly on top of the first exposure (for alignment see Section 3.1.4). The sample is now ready and can be connected to the experiments by wire bonding (Fig. 3.3 (a), step (vi)).

3.2.2 Pattern Transfer with Negative Resists

A negative resist has the characteristic, that only the exposed area remains after the development. This has many advantages, for example, if only a small amount of resist should remain on the sample. A very common negative EBL resist is HSQ ($HSiO_{3/2}$, hydrogen silsesquioxane), which transforms into amorphous SiO after exposure and development, due to breaking of the SiH bonds that are weaker than the SiO_2 bonds [44]. One advantage of HSQ compared to PMMA is the very good minimum feature size of about 10 nm, and very small linewidth fluctuations in the range of 3 nm, possible using EBL [44]. Permalloy nanowires, fabricated with an ordinary lift-off technique have a high edge roughness and possible flagging at the borders. To create wires with smooth edges, a pattern transfer employing HSQ and Ar ion milling was developed. Here, the smooth edges of patterned HSQ, the high resolution of this resist with the low linewidth fluctuations together with the Ar ion milling result in very smooth edges in the patterned material. This is especially useful when edge effects influence the domain wall dynamics. With this patterning method, the edge dependent pinning can be reduced to increase the device performance. A schematic of the process is shown in Fig. 3.7 (b)$^{(i)-(v)}$ and introduced step by step below:

1. HSQ is spin coated on Si/ SiO_2/ Py(25 nm)/ Au(2 nm)/ Cr(20 nm), with Au and Cr acting as a buffer layer for the Ar ion milling process. After coating, the sample is baked on a hot plate for 1 min at 120°C (Fig. 3.3 (b), step (i)).

2. The patterning of 900 nm wide HSQ nanowires is carried out by EBL. The chosen zig-zag shape (see Fig. 3.5) is suited to generate domain walls at the bends. The exposure doses for HSQ are typically of the order of five times higher than for PMMA (Fig. 3.3 (b), step (ii)).

3. After development, the exposed (structurally changed) area remains (Fig. 3.3 (b), step (iii)).

4. The resist is now used as an etching mask for the pattern transfer. Prior to this, optimization for the particular Ar-Ion miller[4] was developed to find the precise parameters that remove the exposed HSQ-Layer and approximately half of the Cr-layer. With this well defined etching rate, only the desired Py wires with a thin Au/Cr capping are left on the substrate (Fig. 3.3 (b), step (iv)).

5. A second step (see Section 3.1.4) is used to generate electric contacts by using steps 1-8 of the list of the pattern transfer with a positive resist in Section 3.2.1. The design and the finished sample fabricated using this method are shown in Fig. 3.5. The contacts right and left are used for current injection into the wire. With these contacts and the additional two contacts at the wire bend, the magnetoresistance of the domain wall can be measured (Fig. 3.3 (b), step (v)).

The samples, prepared with this method are used in the current induced domain wall motion experiment Chapter 5 discussed in *et al.* [9].

3.2.3 Pattern Transfer with Shadow Lithography

Most magnetic materials tend to oxidize, and some samples need to be measured *in situ* without breaking vacuum. In order to solve this problem, shadow lithography is used. An example for a shadow sample is shown in Fig. 3.8. Here a double resist, made of MMA/PMMA was chosen. MMA is a copolymer that is chemically very close to PMMA, but that also slowly dissolves in isopropanol (IPA), which has little affect on PMMA. With this double resist, undercuts of several micrometers are possible and, more importantly, the size of the undercut is tunable. Therefore, with these samples, nano-contacts with well defined notches can be created *in situ*. Previously defined Au-Pads are contacted and connected to the experiment. The

---

[4]The ion milling was performed by Daniel Bedau at Samsung in Korea

3.2. Electron Beam Resists and Pattern Transfer

Figure 3.8: (a) and (b) Scanning electron microscopy image nanocontacts. The PMMA and the MMA is slightly deformed by the SEM electron irradiation. The well defined nanostructure of the shadow evaporated element is visible in the permalloy half-ring on the Si surface. The notch is needed for the experiment and its form size can be tuned by the design of the element. The undercut of several hundreds of nm and a resist hight of above 700 nm in total is sufficient to prevent shortcuts. (c) A schematic drawing shows the different development properties of PMMA and MMA and the resulting undercut, which results in disconnected Permalloy structures.

undercut at the edges of the element, at the wires and the bond-pads is sufficiently large that a short between the metal deposited on the surface of the resist and on the substrate can be avoided. The development of this technique, that generates several samples at a time, replaced a previously used technique that involved several complicated FIB and etching-steps for each element [32]. The process of the sample preparation is as follows:

1. By using steps 1-8 of the pattern transfer with a positive resist in Section 3.2.1, Au contacts are patterned on the chip.

2. A layer of MMA/PMMA was spun on the high resistivity Si sample and baked for one minute at 170°C on a hotplate after each coating step.

3. The nanocontacts are written via e-beam exposure with an overlay procedure (Section 3.1.4) with an overlap to the Au contacts.

4. The resist is developed in a 1:3 MIBK:IPA solution for 90 s in the Hamatech (Fig. 3.8 (c), step (i)).

5. The depth of the undercut can be increased by dissolving some of the MMA in IPA. After approximately 5 min in IPA, the undercut is sufficiently large for the experiment (Fig. 3.8 (c), step (ii)).

6. To check the quality of the undercut, 2 nm Cr is deposited on the resist of a small patterned piece for SEM investigation. If the undercut is not

sufficiently large, the sample is developed again in IPA.

7. After wire bonding the Au contacts, the nanowires can be *in situ* generated by Py evaporation, with the structured PMMA on top acting as a shadow mask (Fig. 3.8 (c), step (ii)).

The samples patterned with this method are focus of the work of J. Heidler and A. v. Bieren and firsts results will be published by A. v. Bieren in [39].

3.2.4 Patterning on Insulating Substrates

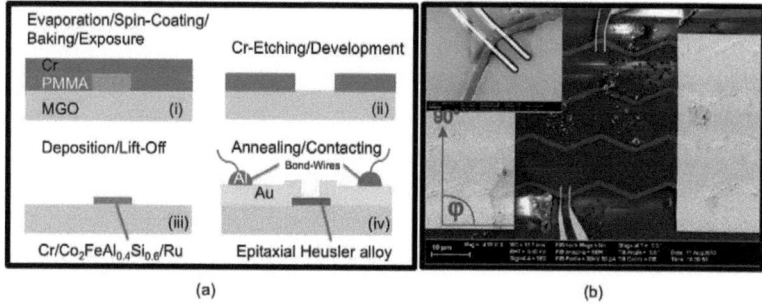

Figure 3.9: The pattern transfer on an insulating substrate is shown in (a). In (i), the MgO substrate with the already exposed PMMA and an additional Cr layer is shown. After the Cr-etching and development, the substrate is opened at the exposed area (ii). The material is deposited and after lift of, the pattern is transferred onto the substrate (iii). An annealing step transforms the alloy into a Heusler compound. An additional overlay procedure is needed to place the electric contacts (iv). In (b), an SEM image of a patterned Heusler alloy on MgO is shown. The contacts (also in the inset) are used to measure the resistance at the wire bend in a four-point geometry.

For many materials or specialized experimental requirements, the magnetic material needs to be grown on insulating substrates. If the sample is insulating, then under an electron or ion beam, charge will build up and the resulting electric field will deflect the ions or electrons, resulting in a reduced resolution or in an unusable exposure. For this reason, patterning on insulating substrates with electrons or ions demands an additional layer of metal on top of the resist to allow the charge to be dissipated. In addition, many insulating substrates, like diamond, MgO or SrTiO$_3$ are transparent, which causes problems with the interferometric height measurement of the EBL system. A recipe for patterning a Heusler alloy (for experiments on Heusler alloys see also Chapter 7), which needs to be grown on transparent and insulating MgO for epitaxially growth, as follows:

1. A 20 nm Cr layer is deposited (Balzer, see Section 3.3) on a PMMA resist that was spin-coated on a [100] orientated MgO substrate. After that, the design is exposed with the EBPG5000Plus. The Cr both conducts the generated charges away and acts as a reflective surface for the interferometric height adjustment of the EBPG5000Plus. Samples written with the Leica Lion LV-1 are capped with only 2 nm Cr to prevent charging and to allow the low energy electrons to expose the resist (Fig. 3.9 (a), step (i)).

2. After removing the Cr layer with a 20-30 s Cr-wet-etch, the PMMA is developed (Fig. 3.9 (a), step (ii)).

3. A $Cr/Co_2FeAl_{0.4}Si_{0.6}/Ru$ multilayer is grown on the sample. After lift-off in acetone, only the patterned elements remain on the insulating substrate (Fig. 3.9 (a), step (iii)).

4. The sample is annealed at 500°C under N_2 atmosphere or in vacuum to recrystallize the thin film into an epitaxial Heusler alloy. To add electric contacts on the structures, the steps 1-3 of this list are repeated with an exposure for the electric contacts and, the bond pads, and Au deposition. After wire bonding, the sample is ready to be used for experiments (Fig. 3.9 (a), step (iv)).

An SEM image of a finished sample is shown in Fig. 3.9 (b). This particular sample is used for transport measurements and measured by J. Heinen. It should be noted that FIB cannot be used for patterning Heusler alloys because this material is very resistant to the ion beam exposure and ferromagnetism remains in our Heusler samples exposed to very high ion doses up to $4000\,\mu C/cm^2$. This is seen in Fig. 3.3, where the stripe domains are not influenced by the patterned triangular shapes.

3.2.5 Patterning of Epitaxial $La_{0.7}Sr_{0.3}MnO_3$ (LSMO)

$La_{0.7}Sr_{0.3}MnO_3$ (LSMO) grows epitaxially on $SrTiO_3$ (STO), when deposited by pulsed laser deposition (PLD) at high temperatures [45]. The temperature required during this deposition method means that the use of a lift-off technique with PMMA is not possible. For the experiment in Chapter 6, structured nano-

and micrometer sized LSMO elements are needed and two different patterning approaches are tested. One technique is FIB lithography and the other one employs a patterned Cr seed layer for controlling the crystalline growth. Experiments on LSMO are presented in Chapter 6. Both methods showed different and interesting effects. Neither of these methods are ideal and the choice of method depends on the experimental needs. Both methods are presented below:

↪ **Patterning of LSMO using FIB**
Starting from a thin LSMO film on STO, the pattern can be directly written by focused ion beam lithography (Section 3.1.3). In Fig. 3.3, a FIB defined area in LSMO is presented. A FIB dose of $100\,\mu C/cm^2$ is not sufficient to remove the magnetic material in LSMO but it is sufficient to change the magnetic properties, from ferromagnetic to paramagnetic. Higher doses are avoided as the gaussian beam shape leads to exposing a larger area to the Ga ions than induced. These effects can be seen in more detailed studies, which are described in Chapter 6.

↪ **Patterning of LSMO using a patterned Cr seed layer**
Epitaxial LSMO is metallic and ferromagnetic and grows epitaxially on STO, but not on Cr. Amorphous LSMO is, in contrast to epitaxial LSMO, insulating and paramagnetic [46]. To make use of the dependence on growth conditions, a 2 nm Cr mask was patterned on a STO substrate (see Section 3.2.4). The 2 nm Cr layer was deposited on the STO and an additional Cr layer was deposited on top of the PMMA. This layer is used to make the transparent substrate reflective, which is required for the interferometric height measurement.

After the exposure and development, the parts of the lower Cr layer not covered by the resist are removed by dry etching. After this step, the PMMA was removed and the LSMO was deposited on the sample. AFM images for a 20 nm and a 60 nm thick LSMO film grown on STO and STO/Cr(2 nm) are shown in Fig. 3.10. On the left side of each image, the LSMO is directly grown on STO and shows epitaxial growth. The LSMO on the right side is grown on Cr and shows amorphous behavior (confirmed by PEEM and SQUID measurements) for the 20 nm thick sample (a) and polycrystalline behavior for the 60 nm thick sample (b). From a set of five samples (20 - 60 nm thick with 10 nm steps), only the 20 nm thick sample grows amorphous on Cr. The other samples showed polycrystalline growth with different grain size that increases with film thickness, implying that the thicker films may allow crystalline growth even on Cr.

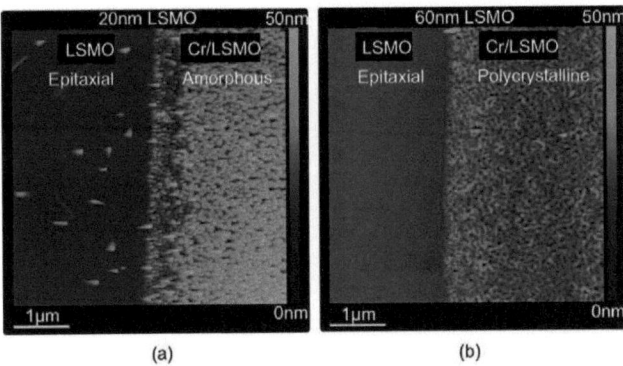

Figure 3.10: (a) AFM image of a 20 nm thick LSMO film patterned epitaxially on SrTiO$_3$ (left) and amorphous on Cr/SrTiO$_3$. (b) AFM image of a 60 nm thick LSMO film, grown epitaxially on SrTiO$_3$ and polycrystalline on Cr/SrTiO$_3$

3.2.6 Patterning on Si$_3$N$_4$ Membranes

Magnetic micro- and nano-structures defined on SiN membranes are used for magnetic measurements involving the transmission of electrons or X-rays such as transmission electron microscopy (TEM) [31] or scanning transmission X-ray microscopy (STXM) [35]. The Si$_3$N$_4$(SiN) membranes are typically 30-200 nm thick in order to keep the absorption as small as possible. The SiN membrane is produced by etching a Si wafer with a continuous SiN layer on top and a SiN etch mask on the back. The membranes are either bought from Silson (U.K.) or made at PSI at the LMN. TEM membranes are typically thinner (~30-50 nm) than STXM membranes (~100-200 nm), as scattering is much higher for electrons than for photons. The structural stability depends on the size of the window and on the thickness of the membrane. Examples of patterned membranes are shown in Fig. 3.11. In (a) and (b) a 200 nm thick membrane for a STXM experiment is shown. In (c) and (d), a TEM membrane is shown. Both samples are designed for current induced domain wall motion experiments. Some STXM experiments are performed at high current densities. The highest possible current density that can be applied is limited by the thermal destruction of the elements, therefore thermal dissipation of joule heating from the sample plays an important role, as this enables the structure to transport the thermal energy due to the ohmic losses into the substrate. The thin membrane is not suited to transport large amounts of thermal energy, which result from the high currents applied through the structure.

To enhance the thermal conductivity and to add an additional thermal reservoir, an additional AlN thin layer is deposited on the membrane. Previously, an Al layer on the back of the membrane was used for thermal dissipation [43] but it was replaced by the insulating AlN on the front of the membrane, since in high frequency experiments, capacitive effects between the element or the current conductor and the underlying metal surface influence the pulse shape and pulse height of the current. In Fig. 3.11 (b), It can be seen that the SiN membrane is buckled due to the mechanical stress on the sample, but these do not influence strongly the structural stability of the sample or the measurement itself. Membranes are

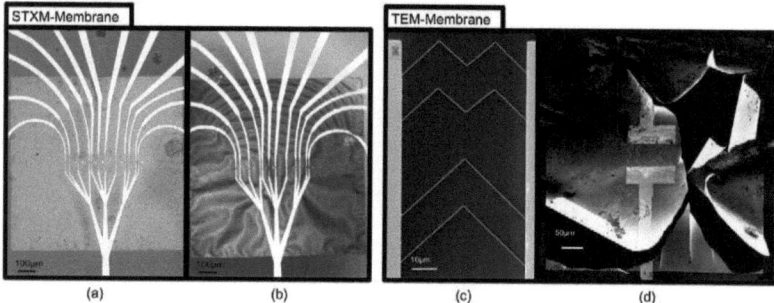

Figure 3.11: (a) Nano- and micrometer sized Py-Wires (500 nm-1.5 µm wide and 8 µm long) on a 1 mm² SiN-membrane are contacted by Au-Wires. The dust on the backside of the membrane is a residuum of the used Protek resist. In later produced membranes, the removal of this resist was optimized. (b) Insulating AlN is put on top of the structures to enhance the thermal dissipation of the joule heating, generated by the ohmic losses during current pulses. The strain, induced by this additional layer is seen in wave like deformations of the membrane. The rigid Au-wired stabilize the membrane a little. These deformations are found no to have no influence on the STXM-measurements. (c) Nanowires on a 50 nm thin TEM-membrane with Au-contacts. The fragile membranes can easily break during lift-off or by wrong handling. A broken membrane is shown in (d).

available for fabrication in two different states; already etched membranes, where the Si below the SiN film is already etched, and unetched wafers with a SiN layer on top and a SiN mask on the bottom which acts a an etching mask. The patterning methods of these two types are outlined below.

Patterning on Etched Membranes

The patterning of nanostructures on membranes is in principe similar to the preparation of patterns on bulk substrates. Only the handling of the membranes and the significantly reduced proximity effect are different. The membrane itself is a SiN-window on a SiN/Si chip with a thickness that can be predefined between a few tens of nanometers up to hundreds of nanometers. These membranes are very fragile, so the use of ultrasounds for the lift-off is not possible and a stirrer in acetone is used instead. Here, the chips are put upside down in the beaker on a special holder [43]. Also, induced stress can destroy a membrane, for example, if a tweezer touches the sample or the sample comes in contact with an uneven surface. A destroyed TEM membrane is shown in Fig. 3.11 (d), in this case, a clamp that was used to keep the membrane in place bent the frame of the membrane a little and the resulting stress accidentally destroyed the membrane. The zig-zag Permalloy wires on a TEM membrane shown in Fig. 3.11 (c) where prepared for M. Eltschka and M. Wötzel and the results of their Lorentz microscopy experiment are published in [31].

Patterning Membranes on un-etched Wafers

We often used un-etched wafers (to allow for the use of ultrasound for the lift-off). Here, the SiN-Layer at the back of the chip is already open but the underlying Si (200 - 300 μm) is not etched. A 50 - 200 nm thick SiN layer on the front-side becomes the membrane window after etching the bulk Si underneath. A single chip for STXM membranes typically consists of a 4×4 array of these 5×5 mm^2 large pre-structured membrane chips. After patterning the samples on the bulk Si-chip with the method described in Section 3.2.1, the front of the chip is protected (Protek^{TM} resist[5]) and the Si is etched using a KOH bath. After removing the protection layer, the membrane array on the chip is ready to be cleaved. A structured membrane, fabricated with this method is shown in Fig. 3.11 (a) and (b). The remover for the protection resist sometimes leave unwanted organic residue on the chip (Fig. 3.11 (a)). This problem was later solved by longer removal times in a special Protec remover followed by proper cleaning with acetone and IPA.

3.3 Deposition of Materials

Many different thin film deposition methods for several purposes and materials exist. The most important techniques for the experiments introduced in this thesis are presented in this section. Thermal evaporation is a standard technique, which is directly implemented at the LMN. For more complicated layers, where,

---

[5]Developed by Brewer Science, Inc.

for example, ultra high vacuum (UHV) deposition or growth at high temperatures is required, the deposition was performed at suited facilities by our collaborators. Py was generally deposited in the molecular beam epitaxy (MBE) chamber of our group[6] at the University of Konstanz (Germany) and at PSI (Switzerland, from late 2010). The Heusler alloy samples where grown in the University of Mainz.[7] Most of the LSMO samples where grown at GREYC (Caen, France)[8] by pulsed laser deposition.

Thermal Evaporation

Single or bi-layers of Cr, Au, Al etc. were deposited in a Balzer BAE 250T system at the LMN. In this technique, the material is placed on a tungsten boat that is mounted on two electrodes. By applying a high current, the tungsten boat with the element heats up due to joule heating and the material is thermally evaporated and deposited on the sample surface that is placed upside down above the evaporator. The film thickness is determined by a quartz thickness monitor and can be confirmed afterwards with a profilometer. The pressure during this process is typically in the range of 10^{-5} mbar, which is high compared to other deposition methods such as molecular beam epitaxy. The film quality is therefore not ideally suited for magnetic materials. Typical uses for Cr/Au layers are the preparation of alignment marks (Cr(6-8 nm)/Au(60 nm)) or for electric contacts to magnetic elements with Cr(6-8 nm)/Au(50-200 nm) with thicknesses that depend on the particular purpose. The Cr acts as an adhesion layer for the Au, which otherwise can be removed easily from the surface.

A metal-deposition MBE system at the University of Konstanz (later at PSI) was used to deposit the Permalloy films. Although Py does not grow epitaxially, the clean vacuum condition of the chamber and the well controlled evaporation parameters lead to reproducibly high quality films with a well defined thickness.

Pulsed Laser Deposition

The $La_{0.7}Sr_{0.3}MnO_3$ films, discussed in Chapter 6, were deposited by pulsed laser deposition (PLD). PLD uses a high intensity laser pulse, to release the target

---

[6]Scientists: J. Heidler, S. Krzyk, A. Patra, A. V. Bieren
[7]Scientists: T. Graf, M. Jourdan
[8]Scientists: L. Méchin, F. Gaucher

material when hit by the focused beam. The material is ablated from a stoichiometric target onto (001) single crystal substrates. The laser radiation energy of the KrF excimer laser (248 nm) was 220 mJ at a repetition rate of 3 Hz; the oxygen pressure was 0.35 mbar and the substrate temperature was held at 720°C during growth. These parameter values were optimized for producing single-crystalline films with smooth surface as judged by X-ray diffraction and atomic force microscopy (AFM). In addition to STO, LSMO can be grown on STO buffered Si chips. This is advantageous for PEEM experiments, where charging of the insulating substrate are an issue.

3.4 Sample Characterization

Before every experiment, and also after every EBL step, the quality and the condition of the sample needs to be checked. In this section, optical and scanning electron microscopy is explained.

3.4.1 Optical Microscopy

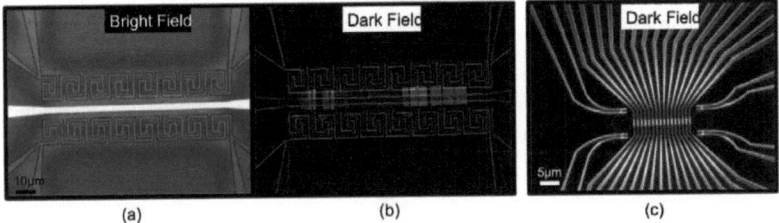

Figure 3.12: In (a) and (b), a bright field and a dark field image from the same sample position are shown. The black area at the patterned region indicates a good development for the micrometer sized area. In (b), nano wires with a line width smaller than the used wavelength are shown in dark field. The absolute success of the exposure cannot be distinguished optically for such small elements. (c) shows an image of patterned nano sized elements. The resolution of the microscope and the angle of incidence does not allow for a characterization of the development status.

The optical microscope has a typical resolution of the order of the wavelength λ of visible light. Nano structures are typically smaller, but for a first impression of the quality of the sample or when other imaging techniques are not feasible, this method is satisfactory, especially as it does not affect electron lithography resists that are still on top of the surface after an exposure. Possible misalignments of an overlay can be determined optically by bright field imaging as shown in Fig. 3.12 (a). Dark field microscopy has the advantage that the success of the development can be determined for micrometer sized elements, as shown in Fig. 3.12 (b). Remaining unwanted resist or other material results in a high contrast, whereas flat areas appear dark. With this imaging technique, areas with residual resist, which were not sufficiently developed, gives rise to optical contrast. Ideally exposed patterns are visible, as well defined bright lines on a dark and flat surface. This method does not work for small structures. In Fig. 3.12 (c), a dark field image of nanowires does not give any additional information about the development status, as the resolution is not sufficient. Also steep and high resist boundaries may result in an artifact in the contrast, so care should be taken with this method when looking at different structures.

3.4.2 Scanning Electron Microscopy

A scanning electron microscope (SEM) has a quoted resolution down to 2 nm and is commonly used in research locations. The image is generated by an electron beam that is scanned over the surface. The generated secondary electrons that exit the sample after bombardment are detected by a detector that measures the total electron yield. Depending on the structure, the images can be taken with an in-lens (IL) detector that sits in the optical axis of the electron beam, resulting in high spatially-resolved top-view images of the sample, or with an secondary electron (SE2) detector. Images taken with the SE2 detector, which sits beside the sample give an indication of the surface if a three dimensional sample is imaged (see for example Fig. 3.8). Further tilting of the sample enhances this effect. The choice of the detector depends on the particular needs.

Due to the electron charge, the sample needs to be conductive. Insulating substrates can be imaged after deposition of a thin Cr layer. To image patterned PMMA, low acceleration voltages and short exposure times are used to prevent permanent damage to the resist. A deformation of an organic resist by the SEM is visible in the PMMA and MMA of the pattern shown in Fig. 3.8 (a) and (b), where a curling of the resist at the edges is observed.

CHAPTER 4

Field Induced Domain Wall Motion

Different concepts have been suggested for next generation solid state memory devices [7, 47]. These memory concepts are based on domain wall propagation, where the wall velocity governs the device performance. Domain walls were predicted to exhibit an inertia-like behavior, which led Döring to introduce the concept of an effective domain wall mass [48]. This is surprising, as the magnetization dynamics can be described using the Landau Lifshitz Gilbert (LLG) equation which, in contrast to the classical mechanical second order differential equation, is a first order differential equation that does not contain intrinsic inertia.

The visible consequences for systems that have mass are the delayed response to an applied generalized force and an oscillatory behavior when relaxing to an equilibrium position. The concept of inertia implies that there is a transfer of energy between different energy reservoirs, as in the transfer of energy between potential and kinetic energies in a mechanical oscillator. In a magnetic system where the energy is transferred between different energy reservoirs, an effective domain wall mass can be seen [49, 50]. However, the nature of the energy reservoirs in magnetic systems has so far not been identified before.

The spin structure of magnetic nano- and micrometer sized element can be influenced in different ways, for instance by a magnetic field. By applying a short intense field pulse, the spin structure can be excited. If this excitation is reversible, for example when the spin configuration has a ground state that is defined by an external field or by its own shape, one can study the change in the spin structure due to the pulse by implementing a pump-probe experiment. The time resolved investigation of field pulse excited domain states was demonstrated by Raabe *et al.* [51]. Here, the high spatial resolution of a photoemission electron microscope (PEEM) is used to image the domain state in a pump and probe manner by synchronizing the excitation of the magnetic state with the orbit clock of the synchrotron. This technique is used here to investigate the dynamic behavior of domain walls in magnetic nanowires. In this chapter, the time and spatially resolved field induced excitations of a single domain wall in a Permalloy half-ring are presented. The data offer new insights into the first nanoseconds of a domain wall displacement, with a spatial resolution of about 50-100 nm and a temporal resolution of ~50 ps. A delay in the onset of the wall motion with respect to the excitation and an oscillatory relaxation of the domain wall back to its equilibrium position was found.

With these results, the origin of both of these inertia effects can be attributed to the transfer of energy between different energy reservoirs based on the exchange and the Zeeman energy. By imaging the distribution of the exchange energy in the domain wall spin structure, these reservoirs were determined, which are the basis of the domain wall mass concept [48]. The main findings, namely, the determination of the energy reservoirs for the exchange and the Zeeman energy, which causes inertia like domain wall behavior as well as the extraction of an effective domain wall mass of $(1.3 \pm 0.1) \times 10^{-24}$ kg, are published in "Imaging of Domain Wall Inertia in Permalloy Half-Ring Nanowires by Time-Resolved Photoemission Electron Microscopy" by Rhensius *et al.* [8].

4.1 Experiment

So far, all measurements of domain wall dynamics in wires have been carried out using transport [50, 52–54] or optical techniques [55, 56]. While these techniques allow for a good time resolution, only the domain wall positions can be determined, i.e., no direct determination of the wall spin configuration during the dynamic processes is possible. On the other hand, various techniques have been used to map the static spin configuration of domain walls [57–59]. In order to understand where the domain wall mass stems from, direct imaging of the dynamic spin structure with high temporal resolution as well as a good spatial resolution is necessary.

4.1.1 Magnetic System

Domain walls in half-ring Permalloy nanowires, where the shape anisotropy makes it possible to tailor the spin structure [57, 58], are an ideal system to study domain wall dynamics (Fig. 4.1). In order to generate a domain wall at a well defined position of the ring, a strong magnetic field (B_{Init}) is applied (for field direction see Fig. 4.1 (a)). This field now aligns all the spins of the element parallel to B_{Init}, schematically shown in Fig. 4.1 (a). By reducing this field to zero, the spins on the right and on the left side of the ring point towards each other, as they follow the ring shape in order to reduce the stray field energy (shape anisotropy). The resulting domain state, is shown in Fig. 4.1 (b), where the arrows indicate the resulting spin direction.

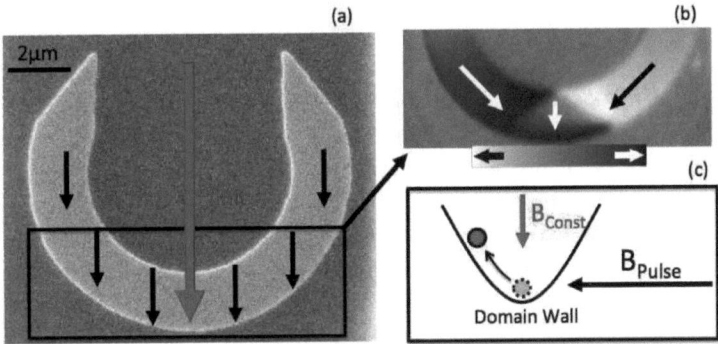

Figure 4.1: (a) SEM image of a Permalloy half-ring structure. The red arrow indicates the strong field, used to saturate the spins in the structure (black arrows, oriented parallel to B_{Init}). An XMCD image taken at remanence after magnetic saturation is shown in (b). The arrows indicate schematically the spin direction (a detailed image of the spin structure is given in Fig. 4.2 (b)). In (c), the potential generated by a small constant field B_{Const} parallel to the initializing field B_{Init} is schematically shown. The direction of the pulsed field points to the left and forces the domain wall (red dot) to a new position, resulting in an increases of the domain wall energy.

A small permanent field B_{Const} of 4 mT that points in the same direction as B_{Init} generates a potential for the domain wall, which is harmonic to first order of approximation for small displacements. To excite the domain wall, a short field pulse perpendicular to the constant magnetic field is applied. Fig. 4.1 (c) shows schematically the working principle of the experiment. The domain wall is symbolized by the red circle.

In this experiment, the interest is to understand how the domain wall responds to the field pulse and how it moves back to its equilibrium position. In the initial conditions of the experiment, an asymmetric transverse domain wall [57] at the bottom of the ring element was observed, which is trapped at this position by the applied constant magnetic field B_{const}. This domain wall consists of a central triangular domain (TD) with homogeneous magnetization pointing down and two domain boundaries, one on the left (B_L) and one on the right (B_R) side of the TD (see Fig. 4.2). The spins rotate between the triangular domain and the region at the bottom structure edge within an area denoted as (B_E). For the observed sample dimensions, one would usually expect vortex walls instead of transverse walls [58]. The permanent applied field shifts the phase diagram in a way that transverse walls are possible in wider elements (this behavior is in line with a detailed study about domain wall types in LSMO ring elements at different applied magnetic fields, described in Chapter 6). Another effect that supports transverse domain walls after magnetizing is pinning, which may suppress the formation of a vortex wall (a more detailed study about the formation of a vortex wall from a transverse wall can be found in Chapter 6.4). By visualizing the exchange energy (Fig. 4.7 (a)) which is proportional to $(\nabla M)^2$, one can see a region with a large exchange energy at the intersection of B_E and B_R. This might be an indication of a local pinning site. The domain wall type is not crucial for this particular experiment as all domain wall masses are measured for both vortex, and transverse walls. However, different behavior is expected for transverse and vortex domain walls. This can result in different depinning fields and also in different domain wall masses [50].

4.1.2 Samples

The sample consists of Py rings, grown on top of a Au stripline. The gold stripline, with a width of 10 μm and a thickness of 200 nm, is fabricated by optical lithography. The Au surface for the Py deposition should be as flat as possible to reduce unnecessarily induced pinning sites. For smooth Au layers, a DC magnetron sputtering system was used.[1] The half-ring nanowires are fabricated on top of the stripline by electron beam lithography (Leica LV1) combined with a lift off process [60] (see also Section 3.2.1). The main results were obtained on a structure with a diameter of 6 μm and a width of 2 μm. The overlay was performed by using crosses, which are defined in the optical lithography step used to manufacture

---

[1] The DC magnetron sputtering was performed by Michael Horisberger at PSI.

4.1. Experiment

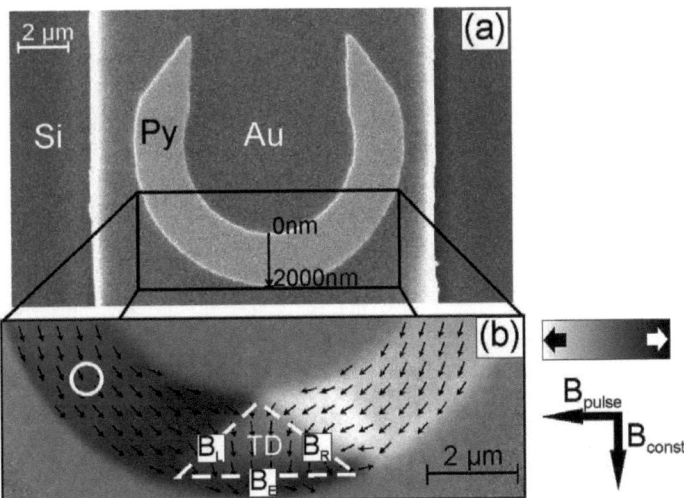

Figure 4.2: (a) Scanning electron microscopy image of a Permalloy half-ring nanowire placed on a gold stripline and (b) corresponding XMCD image of an asymmetric transverse domain wall, where the contrast given by the gray-scale bar corresponds to the magnetization direction as visualized by the black arrows.

the stripline. The half-ring material is 20 nm of Permalloy, deposited by MBE (see Section 3.3) with a 1.5 nm Au capping to prevent oxidation. A close up SEM image of the element on the Au stripline is shown in Fig. 4.2 (a). A larger overview of the stripline is given in Fig. 4.3 (a). For the experiment, high current densities in the stripline are required. Therefore, very high ohmic Si-substrates were used to prevent possible conduction through the silicon.

4.1.3 Experimental Technique

By combining X-ray magnetic circular dichroism photoemission electron microscopy (XMCD-PEEM) imaging with a pump-probe technique, the domain wall spin structure can be imaged as a function of time. The high time resolution of the XMCD-PEEM is achieved by triggering the detector on the single 70 ps wide X-ray pulse of the camshaft bunch with a variable time shift with respect to the field excitation [51] (for Synchrotron light see Section 2.1). In order to create a pulsed in-plane magnetic field, a laser pulse with a pulse width of 15 ps and a pulse repetition rate of 63 MHz, which is synchronized with the synchrotron beam, is used to trigger a photodiode, which causes an electrical current to pass through a stripline (Fig. 4.3 (b)). A schematic of this setup is given in Fig. 4.3. The current, which flows through the stripline after the laser hits the photodiode, is indicated. The resulting pulse shape (Fig. 4.5 (d) (black curve)) is extracted from the total electron yield during the field pulse. The total electron yield changes with the stripline current, as the photoelectrons with a kinetic energy in the range of 1-2 eV can be influenced by an applied voltage. An image of the setup can also be seen in Fig. 2.2 in Chapter 2.

A field pulse amplitude of 3.5 mT is reached after a very short rise time of 150 ps (Fig. 4.3 (c)). This value is determined by taking the total current and the field pulse shape into account. The field pulse has a width of less than 500 ps and the complete pulse decays after 1800 ps. The pulse in the central region of the stripline is mostly in-plane, and the out-of-plane component at the edges can be neglected [61]. In order to optimize the field pulse, different laser intensities (resulting in different field pulses) were tested together with different permanent magnetic fields. Unfortunately, it was found that only a small operating window for excitation and permanent displacement, where the domain wall moves too far and cannot be reinitialized by the constant field, exists. After increasing the laser intensity beyond a critical value, the domain was permanently displaced and needed to be reinitialized with B_{Init}. The experiment was therefore performed at a laser intensity, that is just below the critical value that is needed for an irreversible displacement in order to maximize the absolute displacement.

The field B_{Const} that is used to restore the magnetic domain wall back to its equilibrium position was found to be ideal at 4 mT. It is permanently applied using permanent magnets located underneath the sample (for field directions, see Fig. 4.2 and Fig. 4.3 (d)). To vary the field, several combinations of small permanent magnets with pole shoes were glued on different Al plates. By adding Mu-metal pieces on the side of these magnets, the permanent field can be reduced in very fine and well controlled steps. Several permanent magnets were built with field values between 2 mT and 10 mT. The field of every magnet combination was measured with a Hall probe at the position were the structure is placed in the experiment. The optimal permanent magnetic field was determined experimentally. Therefore, the aluminum plates with the mounted magnets were exchanged until a field was found, sufficient to restore the domain wall reproducibly and small enough to allow for an observable domain wall displacement during the field pulse.

4.2. Results and Discussion

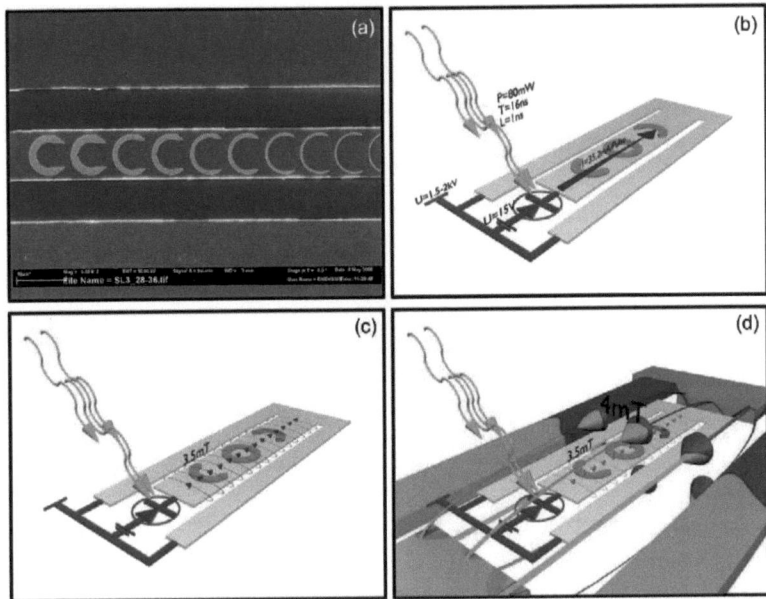

Figure 4.3: (a) SEM image of the Au-stripline with Permalloy half-ring elements on top. The coplanar waveguide geometry (b) is connected to a photodiode, which is biased with 15 V to be triggered by a nanosecond laser pulse. The generated charge flows through the stripline producing a total current of 35.2 mA. (c) Shows the generated field profile, with a peak of 3.5 mT, which is mostly in-plane on the stripline surface. (d) The constant field B_{Const} is generated by two permanent magnets, coupled with two pole shoes. This set-up is placed below the circuit board and the stray field in the center is mostly in-plane and points perpendicular to the field that is generated by the stripline.

4.2 Results and Discussion

Images are taken at various moments in time with respect to the field pulse. The field pulse starts at a relative time shift of 11.05 ns and the shift is increased in steps of 50 ps. This is less than the predicted temporal resolution of 70 ps, but time resolved changes in the spin structure down to time steps of 25 ps are observed, which means that the effective temporal resolution is below 70 ps. In Fig. 4.4 an image of the wall position just before the beginning of domain wall motion (Fig. 4.4 (a)) and during the field pulse (Fig. 4.4 (b)) with a time difference of 200 ps is shown. The wall (B_L) is shifted by 440 nm to the left (compare stars in Figs. 4.4 (a) and Fig. 4.4 (b)), yielding a very high velocity of 2200 m/s for this

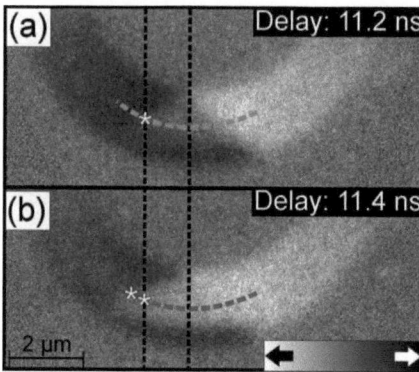

Figure 4.4: Spin configuration in the asymmetric transverse domain wall (a) during the field pulse, just before the domain wall starts to move, corresponding to the time shift 11.2 ns, and (b) 200 ps later, where B_L is displaced by 440 nm.

part of the wall. Such high speed is in line with earlier measurements [55]. Slower average velocities obtained in transport measurements [7] are due to the fact that the velocity is limited at long time scales by the Walker breakdown [62].

In order to analyze the domain wall during the displacement, several spatially averaged line profiles (parallel to the green and red dashed lines in Fig. 4.4 (a) and Fig. 4.4 (b)) across the nanowire (Fig. 4.5 (a)) are extracted. The 2000 nm wide nanowire is divided into 10 spatially averaged slices, each 200 nm wide. For every slice, line scans are extracted for 55 consecutive time steps separated by 50 ps, making it possible to observe the time evolution of the spin configuration in each slice over 2750 ps (see Fig. 4.5 (a)). Comparing all of the slices, one can see that the time evolution of the magnetization varies from slice to slice, so demonstrating magnetization dynamics that can only be revealed with direct spatially resolved imaging. In addition, our experiment compares well with micromagnetic simulations (Fig. 4.5 (b)) [17]. For the simulation, the same nanowire dimensions and field parameters employed in the experiment are used, and standard parameters for Permalloy are used: a damping $\alpha = 0.02$, a saturation magnetization of $M_S = 800\,\text{kA/m}$, an exchange constant of 13 pJ/m, a gyromagnetic ratio $\gamma = 2.2 \times 10^5$ m/As, and a lateral cell size of 5 nm. As a guide to the eye, dashed orange lines mark the position of the wall before the pulse and the position of maximum displacement and, since they are not parallel, there is a nonuniform displacement of the left-hand boundary, leading to a change of the spin structure of the domain wall during the displacement.

4.2. Results and Discussion

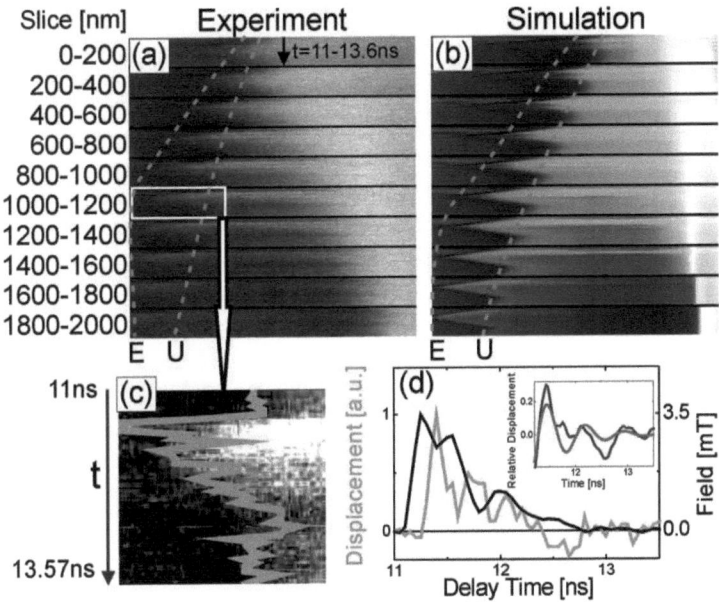

Figure 4.5: Time evolution of the domain wall dynamics (a) in the experiment and (b) in the simulation. The ring shaped structure is divided into ten 200 nm slices and the average contrast for each slice is given as a function of time for t = 11 to 13.57 ns. The dashed orange lines indicate the left boundary (B_L) of the wall in the ground state (U = unexcited) and at the highest wall displacement (E = excited) (compare with Fig. 4.2). (c) Time evolution of B_L in the slice 1000-1200 nm. The region corresponds to that indicated by the white frame in Fig. 4.5 (a) and the wall position deduced from the line scans is indicated in green. (d) The domain wall displacement taken from Fig. 4.5 (c) (green) together with the applied field pulse (black). The inset shows the displacement, after subtraction of the running average, yielding the oscillation frequency. The red curve is a damped sine fit used to determine the oscillation frequency.

From the slice 1000-1200 nm in Fig. 4.5 (a), the position of B_L is extracted as a function of time (shown in green in Fig. 4.5 (c)). This curve is shown as a function of time in Fig. 4.5 (d) (green line) together with the time evolution of the field pulse (black line). A surprising behavior is observed: first, it is only at 200 ps after the start of the field pulse that the domain wall starts to move in the field direction and reaches its maximum displacement 150 ps after the maximum in the field pulse. Second, as the field pulse decays, the wall moves back due to the permanently applied field, and it undergoes a damped oscillation around its equilibrium position (Fig. 4.5 (d)). Using a method of subtracting the running average as detailed in Ref. [51] (inset of Fig. 4.5 (d)) the oscillation frequency is deduced to be 1.3 ± 0.6 GHz.

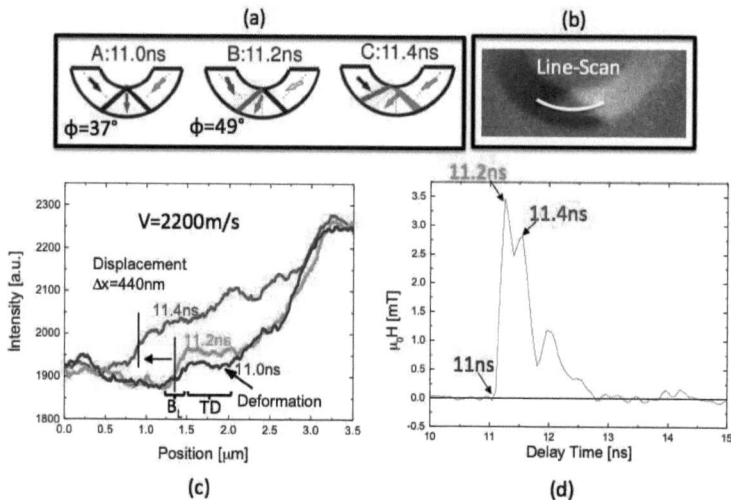

Figure 4.6: (a) Schematic of the domain wall excitation before the field pulse at 11.0 ns, the excited domain at 11.2 ns, and the displaced domain wall at 11.4 ns. (b) The line indicates the position of an intensity line scan that crosses B_L and B_R. Line scans for 11.0 ns, 11.2 ns, and 11.4 ns are shown in (c): the blue line is the ground state before the pulse is injected into the stripline at 11.0 ns. The gray contrast in the center of the domain wall (TD) is visible in the small plateau. B_R is the steep rising flank at the right side, B_L is the flank on the left side. The green line shows the excited domain wall at 11.2 ns. The red curve shows the contrast of the line after the domain wall depins at the left side. The wall moves ~440 nm within 200 ps to the left, which means a very high velocity of around 2200 m/s. The field pulse, labeled with the particular delay time positions is shown in (d).

These two observed phenomena, the delayed displacement and the oscillation, are both consequences of what is usually described as inertia. The transfer of energy between different energy reservoirs leads to this inertia-like behavior and domain wall mass. The domain wall at three relevant times is schematically shown in Fig. 4.6 (a). Prior to the field pulse at a delay time of 11.0 ns (A), the spins make an angle of $\sim 37°$ on the left side of the unexcited ground state of the domain wall (B_L). The excited domain wall at 11.2 ns (B) has not moved yet, but

4.2. Results and Discussion

the angle at B_L increased to $\sim 49°$. The spins in the three mapped regions rotate inhomogeneously, which leads to an increased exchange energy that is visualized by a red wall in Fig 4.7 (a). 200 ps later at a delay of 11.4 ns (C), the displacement of B_L to the left is shown. At this point, the Zeeman energy is at its maximum. In order to identify these energy reservoirs in our experimentally observed magnetic system, intensity profiles across the wall (see Fig. 4.6 (b)) for the time before the field pulse (11.0 ns, see Fig. 4.6 (c) blue) and during the field pulse just before the wall starts to move (11.2 ns see Fig. 4.6 (c) green) are plotted. The red curve is the line scan of the displaced domain wall.

By comparing the blue and green intensity profiles in Fig. 4.6 (c) one can see that the position of B_L has not changed in the first 200 ps of the field pulse. The plateau TD increases, which means that the spins have rotated in this region and that the angle of the spins across B_L increases. The angle between the spins to the left and the right of B_L is increased from $37°$ (blue intensity profile Fig. 4.6 (c)) to $49°$ (green intensity profile Fig. 4.6 (c)). To understand how this change in the profile of B_L is related to the delay in the displacement, one has to consider what the effect of the applied field is for the system: as soon as the field pulse is switched on, it generates Zeeman energy, which is then expected to lead to the wall displacement. What is found, though, is that it first leads to a change in the profile of B_L, which changes the exchange energy stored in this region. The increased magnetization gradient inside B_L (green profile in Fig. 4.6 (c)) means that Zeeman energy is transferred into exchange energy and thus the magnetization of B_L acts as an energy reservoir, which is the prerequisite for this inertia-like delay behavior.

To fully analyze this energy transfer, the spatial change of exchange energy during the pulse excitation is imaged. Since the exchange energy is $E_{ex} \propto (\nabla M)^2$, the squared gradient of the image intensity is proportional to the spatial distribution of the exchange energy inside the domain wall. The different states of the excited domain wall are schematically shown in Fig. 4.7 (a). An image of this distribution is shown in Fig. 4.7 (b). In this image, the local change of the XMCD contrast was used to map the exchange energy of the ground state of the system shortly before the excitation in order to compare this energy distribution with the one of the excited state. The areas B_R, B_L and B_E have high exchange energy, which is marked with red ovals. At the overlap of B_E and B_R, very large variations for the exchange energy are observed. This might be due to the fact that, in the half anti-vortex at this position, there is a large rotation of the spins at small distance. It could also be that the domain wall is pinned at that position resulting in a large angular variation of the spins close to the pinning site. Both

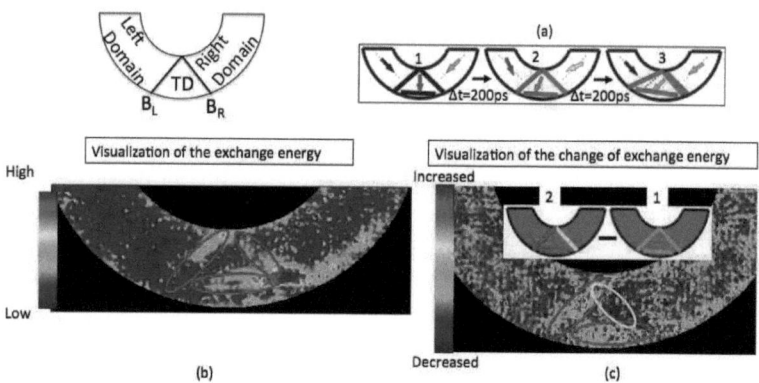

Figure 4.7: (a) The net change of the exchange energy of the domain wall is shown schematically in 200 ps time intervals. In **1**, the domain is at rest. There is no change of the exchange energy. The position of B_L and B_R is marked in black. In **2**, 200 ps later, the spins of the region TD are rotated more, than the spins in the domain right and left of the domain wall. The total exchange energy in B_L and B_E increases (marked red) and the exchange energy in B_R is reduced (marked blue). In **3**, the excited domain wall is displaced. The local distribution of the exchange energy of our observed domain wall at rest is shown in (b). By mapping the exchange energy at different times, the local changes of this energy can be highlighted by subtracting images at different times. In (c), the difference of **2 - 1** is shown. One can clearly observe an increase in exchange energy at B_L and B_E (red), and a decrease at B_R (blue).

could cause this large contrast.

By performing the same calculation for the spatially mapped exchange energy for the excited state 200 ps later, one can directly compare these time resolved distributions. This is shown in Fig. 4.7 (c). At the position of B_L and B_E (marked by the two red ovals) one can see that the exchange energy increases. Therefore the magnetization in B_L and B_E act as energy reservoirs. Due to the spin rotation in the right domain, the exchange energy in B_R decreases, observed as dark blue (see yellow oval in Fig. 4.7 (c)). For these areas, the change in exchange energy stems from the fact that the spins inside the TD rotate uniformly as they are subjected to the largest torque due to their perpendicular orientation with respect to the field pulse direction. The spins to the left of B_L and at the edge below B_E do not rotate as much, so that the total angle and thus the gradient across B_L and B_E increases, whereas it decreases across B_R. The areas with uniform spin rotation do not show a change of the exchange energy, which can be seen in the green-turquoise contrast of TD (i.e. the lack of red-yellow contrast) and the areas left

4.2. Results and Discussion

and right of the domain wall (right and left domain).

To further underpin this key finding, micromagnetic simulations are performed to determine the parameters that govern the energy transfer and thus the inertia. Most important is the exchange constant, which governs the efficiency of the transfer from Zeeman to exchange energy. To show the influence on the inertia, the exchange constant was varied and the resulting delay and oscillation frequency was determined. The results of the simulation are plotted in Fig. 4.8.

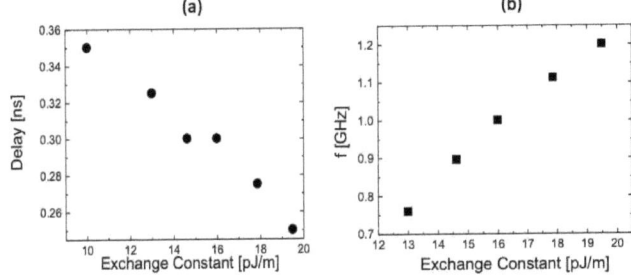

Figure 4.8: The dependence of the delay between the field pulse and the wall displacement (a) and of the oscillation frequency (b) on the exchange constant as deduced from micromagnetic simulations are shown.

First, a decreasing delay between the field pulse and the onset of the wall motion was found with increasing exchange constant as shown in Fig. 4.8 (a). To understand this, the critical wall angle (angle between the spins across B_L and B_E) was determined at the moment that the wall starts moving. It was found that this wall angle is smaller for larger exchange constant. The explanation for this behavior is that, in order to trigger the wall motion, a critical torque is necessary and, for a given angle between any two spins (magnetization gradient across the wall), this torque scales with the exchange constant. Therefore, with increasing exchange constant this critical torque is reached sooner when the angle between the spins is small (i.e. smaller magnetization gradients).

The second visible consequence of inertia is the behavior of the oscillation frequency of the domain wall, as it relaxes around its equilibrium position. In Fig. 4.8 (b) an increase of the oscillation frequency with increasing exchange constant is shown. To analyze this, the formula for the oscillation frequency of a domain wall in a curved wire is considered [47], we expect:

$$f \propto \frac{1}{\sqrt{m}}, \tag{4.1}$$

where f is the oscillation frequency and m is the domain wall mass. Since the mass m depends on the domain wall width λ [47]:

$$m \propto \frac{1}{\lambda^2} \qquad (4.2)$$

and

$$\lambda = \sqrt{\frac{A_{ex}}{K}} \qquad (4.3)$$

where K is the anisotropy, in our case due to shape, A_{ex} = exchange constant, and $f \propto \sqrt{A_{ex}}$ is expected. The dependence of an increasing f with increasing A_{ex} agrees with our simulations (Fig. 4.8 (b)).

Both the delay and the oscillation frequency depend strongly on the exchange constant, which is thus the governing parameter for the inertia. To obtain a quantitative result of the domain wall mass, the oscillation frequency f is determined to be 1.3 ± 0.6 GHz. Its mass can be deduced using $f^2 = \frac{Q_m H}{4\pi^2 mr}$ where the applied field is H=4 mT, the radius of the structure $r = 3\,\mu$m and $Q_m = 2\mu_0 M_s S$ is the magnetic charge in zero field with the saturation magnetization M_s=860 kA/m and the cross-sectional area $S = (2 \times 10^6) \times (20 \times 10^9)$ m^2. The calculated mass is $(1.3 \pm 0.1) \times 10^{-24}$ kg [49], in line with previously measured values [49, 50, 52].

To check that our observed inertia behavior is related indeed to the inhomogeneous wall spin structure and not an artifact of the measurement, the spin dynamics away from the wall in the region indicated by the circle in Fig. 4.2 (b) is analyzed. The spins here follow the field almost instantaneously as expected from the LLG equation for a macrospin. The frequency of this oscillation is 3.4 ± 0.6 GHz, which corresponds to a precession of the spins around the effective field. From the micromagnetic simulations, the strength of the effective field was determined to be 114 kA/m. The oscillation frequency can be determined by the Kittel formula [63]:

$$\omega = |\gamma| \sqrt{\mu_0^2 \left(H_{Bias} + H_{Ani}\right)\left(M_s + H_{Bias} + H_{Ani}\right)}, \qquad (4.4)$$

with μ_0 the vacuum permeability, H_{Bias} the applied field and H_{Ani} the acting anisotropy (in our case shape anisotropy) and γ the gyromagnetic ratio. An expected precession frequency of $\sim$4 GHz is obtained, which agrees very well with the experimentally observed value.

4.3 Conclusion

In this experiment, the domain wall displacement in a Permalloy nanowire with a temporal resolution in the range of 50 ps was imaged. A very high domain wall velocity of above 2000 m/s was observed. This velocity is not limited by the Walker breakdown, due to the short time scale. Inertia-like behavior was found, visible in both a delay in the onset of the domain wall motion to the applied magnetic field pulse and in an oscillatory behavior as the domain wall returns to its equilibrium position. From the tempo-spatially resolved results it was concluded that the domain wall inertia stems from the energy transfer between two energy reservoirs, the exchange energy inside the domain wall spin structure and the Zeeman energy of the wall in the magnetic field. These energy reservoirs are equivalent to a classical system as illustrated in Fig. 4.9. The potential energy of a deflected pendulum is analogous to the Zeeman energy of a displaced domain wall in the attractive potential of the constantly applied field. However, the kinetic energy of a mass swinging back does not exist in a magnetic system. In the case of a moving domain wall, it was shown that the exchange energy is increased, resulting in a deformation of the wall, which can be compared, for example, to the deformation of a bouncy ball. These two energy reservoirs (namely the exchange and the Zeeman energy) and the energy transfer between them now provides a system that shows inertia-like behavior, which was successfully imaged. During the first 200 ps time interval, when the field pulse is applied but the domain wall is not displaced, the domain wall boundary becomes distorted and so stores exchange energy. This was imaged by subtracting the contrast of the domain wall at different excitation states and by comparing local time resolved line scans across the wall. This deformation and therefore accumulation of exchange energy, leads to the delay between the field pulse excitation and the onset of wall motion. The damped oscillation, as the wall relaxes to its original position can also be attributed to the transfer of energy between these reservoirs. Micromagnetic simulations and analytical models indicate that the inertia and the resulting domain wall mass is governed by the exchange constant.

With the information gained from the domain wall oscillation, the effective mass of this particular domain wall was determined to be $(1.3 \pm 0.1) \times 10^{-24}$ kg, which is in line with other investigations of domain wall masses [49, 50, 52].

This effective domain wall mass and the related inertia of the domain wall is of particular scientific and technological interest, as it limits the onset of the domain wall motion and has a strong influence on the resulting domain wall dynamics.

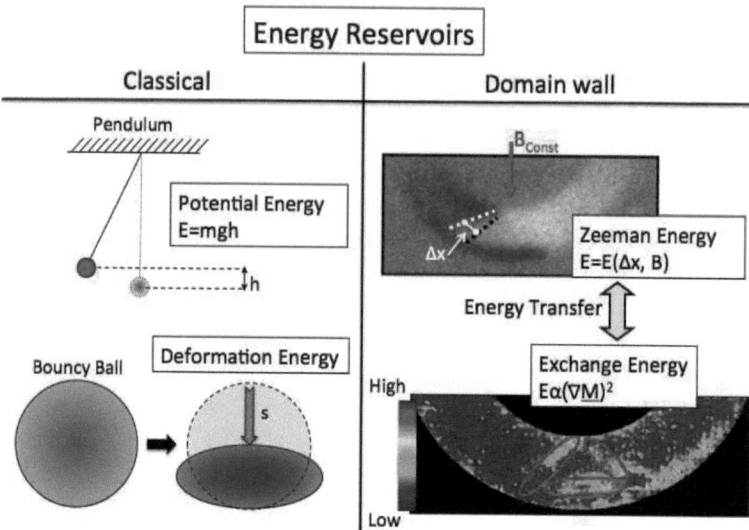

Figure 4.9: Possible energy reservoirs of a ferromagnetic system (in this case a domain wall) are shown and compared with classical energies. The potential energy of a pendulum is comparable with the displaced domain wall in a constant field. The exchange energy is compared with the compression of a bouncy ball. In both cases, the x-position does not change, but the energy of the system is increased by squeezing or deforming local states.

The time resolved images show this influence for the first time in a spatially resolved manner and therefore give a more detailed understanding, which could not be achieved in earlier experiments [49].

CHAPTER 5

Current-Induced Domain Wall Motion

The possibility to move a magnetic domain wall with electric current instead of a magnetic field is of high scientific interest and opens the door to a wide range of applications. The force on the domain wall generated by the spin torque [20, 64] (see Section 1.3.2) pushes the domain wall in the direction of the electron flow, which leads to a displacement of the wall when a critical current density is reached [65–68]. The advantage compared to field-induced domain wall motion is that the domain walls move in a direction that only depends on the current direction and not on the spin configuration. A chain of domain walls in a nanowire can now be displaced at the same time, which could be used for mass media storage devices with large storage density. Such a device is called racetrack memory as proposed by Parkin [7]. Even more importantly, no fast mechanically spinning parts are involved, such as those in state of the art magnetic hard drive devices. However, stochastic pinning sites and thermally activated processes are not yet well controlled and therefore the domain wall displacement is not absolutely reproducible, which is a prerequisite for technical applications.

Recent studies on *in situ* domain wall motion concentrate on the fundamental processes that govern the spin torque and the domain wall velocities [25, 33, 36]. The device performance is proportional to the domain wall velocity and velocities above 100 m/s induced by ultra short current pulses have been reported [25].

The spin torque efficiency scales with P/M_s [25], therefore materials with a high spin polarization P at the Fermi level, and a low saturation magnetization M_s are ideally suited for current-induced domain wall motion. A material commonly used is Permalloy, as the saturation magnetization is about $800\,\text{kA/m}$, the spin polarization at room temperature is around 40% [69] and it has a very small magnetocrystalline anisotropy.

However, the high current densities in the range of $10^{12}\,\text{A/m}^2$ that are needed for this material are problematic for both experiments and applications. Firstly, a high current results in a very high power consumption, undesirable for applications and secondly, the Joule heating and electromigration can even destroy the nanowire. Joule heating can also affect the local spin configuration by heat assisted nucleating or annihilating domain walls [70, 71] and give rise to domain wall transformations [71, 72] that are also observed in structured ferromagnetic half-metals e.g. LSMO and Heusler alloys (see Chapter 6 and Chapter 7). The critical current density also depends on the applied magnetic field, the domain wall type, the wire width, and the local pinning sites of the element. Furthermore, the pinning of the domain wall in a nanowire is influenced by the edge roughness of the nanowire [9].

The first part of this Chapter discusses results on current-induced domain wall motion in Permalloy nanowires that have less edge roughness than Permalloy wires fabricated by ordinary lift-off techniques. The smallest current density needed to push the domain wall in the current direction is the critical current density j_c, which was found to be four times smaller than the critical current densities in similar geometries patterned with lift-off. The linear dependence of the depinning field on the critical current density was demonstrated and at current densities above the critical current density, domain wall transformations have been observed.

The second part discusses time resolved current induced domain wall dynamics. A pump-probe PEEM experiment was used to image the onset of the current induced domain wall motion *in situ*. Vortex domain wall displacements have been observed and the measured velocities are in line with the results reported by L. Heyne *et al.* [25].

Part of the results of this chapter have been published by Grégory Malinowski *et al.* [9].

5.1 Displacement in Materials with low Depinning-Field

This section discusses the results on domain wall displacement of nanowires made by ion milling with an HSQ mask. This special preparation method was chosen in order to generate nanowires with smooth edges, which reduces the influence of edge defects. The clue about good edge conditions is based on high resolution HSQ resist that shows low linewidth fluctuations [44], combined with a subsequent ion milling pattern transfer. A detailed description of the patterning method used is presented in Section 3.2.2. The wires, produced with this patterning method have 4-5 times lower depinning fields than wires of the same material made by ordinary lift-off technique [73].

Both the critical current density for *in situ* domain wall motion and the depinning field of a domain wall are influenced by the edge roughness [9].

5.1.1 Experiment

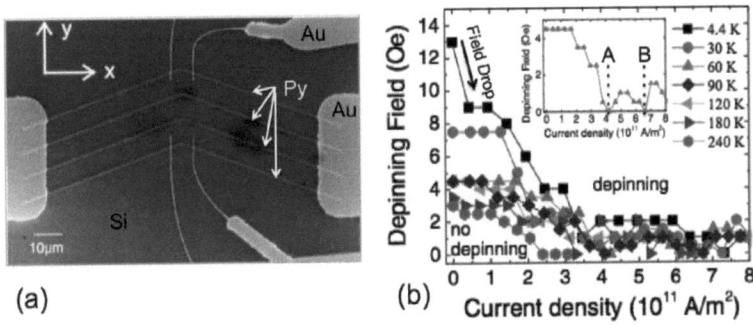

Figure 5.1: (a) SEM image of the contacted 900 nm wide and 25 nm thick zig-zag nanowires. The depinning fields of the domain walls as a function of current density are shown in (b). The direction of the forces that act on the domain walls are chosen to be equal for the field and the current. The used current pulse has a constant length of 50 μs. The inset shows the dependence at a cryostat temperature of 60 K.

An SEM image of a typical sample is shown in Fig. 5.1 (a). Two Au-contacts, one on the right and one on the left contact the 900 nm wide and 25 nm thick Permalloy nanowires. The top and bottom wire have two additionally Au-contacts, one to left and one to right of the kink, that are used to keep track of the domain state between these contacts by measuring the magnetoresistance in a four point geometry. Prior to the experiment, the domain wall needs to be initialized at the bend of the wire. To reproducibly set the domain wall position, a strong magnetic field is applied perpendicular to the wire (y-axis in Fig. 5.1 (a)), similar to the field-induced experiment in Chapter 4. After switching off the field, the spins right and left of the bend point anti-parallel to each other and generate a head-to-head or tail-to-tail domain wall, depending on the field direction. For this wire geometry, vortex walls are the stable domain wall spin configuration [58].

To assist the *in situ* domain wall motion, a field in the direction of the current flow is applied. This field is increased for a given current density, until a domain wall displacement is observed. The critical depinning current density for each field value (applied in the x-direction) is determined by applying 50 μs current pulses with increasing amplitude. After each pulse injection, the DW is reinitialized by a field. The existence of a domain wall at the bend is determined by the AMR-effect in a 4-point geometry which is shown in Fig. 5.1. The resistance is lowered when a domain wall is present between the Au-contacts and increases when the domain wall leaves this region. The depinning measurements were performed in a bath cryostat at temperatures in the range of 4.4-240 K.

5.1.2 Results

In Fig. 5.1 (b), the depinning field as a function of the current density is shown. At zero current, the depinning clearly depends on the sample temperature. The necessary field is reduced from 13 to 3 Oe by increasing the temperature from 4.4 to 240 K. This dependence shows that the depinning is driven by thermally activated processes. These depinning fields are about three times smaller than in similar samples prepared by a standard lift-off technique [73]. This fact already indicates the effect of the edge roughness on the depinning fields.

For small injected currents up to $\sim 1 \times 10^{11}\,\text{Am}^{-2}$ the depinning field remains constant, with the exception of a reduction at a cryostat temperature of 4.4 K. This effect might be due to Joule heating, previously observed in [74]. For current densities above $1 \times 10^{11}\,\text{Am}^{-2}$, the necessary field for a domain wall displacement decreases approximately linearly with the current until the critical current density $j_c(H = 0)$ is reached. The measured critical temperature $T_c^{H=0}$ for cryostat temperatures below 100 K seems to be independent of the temperature at values around $4 \times 10^{11}\,\text{Am}^{-2}$.

5.1. Displacement in Materials with low Depinning-Field

After further increasing the current density, the necessary field also increases to values > 0 Oe for temperatures below 240 K (apparent for instance in the inset of Fig. 5.1 for current densities between ~ 4 and $\sim 7 \times 10^{11}$ Am^{-2}, marked with 'A' and 'B'). This behavior can be explained by domain wall transitions that occur during the displacement that lead to higher depinning fields. A reliable and temperature dependent depinning is achieved for current densities in the range of $\sim 7 \times 10^{11}$ Am^{-2}. The strong temperature dependent depinning requires a correction of the sample temperature, which is not equal to the bath temperature due to Joule heating [75].

The dependence of j_c on T is shown in Fig. 5.2 and shows 5 times smaller critical current densities and higher pinning fields than samples prepared by lift-off. The critical current densities decrease from ~ 4 to $\sim 2.5 \times 10^{11}$ m^{-2} with increasing sample temperature.

The comparison with 1500 nm wide Permalloy wires prepared by standard lift-off technique and a similar thickness demonstrate the relative reduction of the depinning field in this case. Wider wires have lower depinning fields (see Chapter 6), of about ~ 10 Oe [73] at room temperature, which is $\sim 4-5$ times larger than the Ar ion milled samples. In addition, the critical current densities in ordinary Permalloy wires, typically around $\sim 1.3 \times 10^{12}$ Am^{-2}, are also 5 times higher than observed in the Ar ion milled sample. These findings show a clear dependence of the depinning field on the edge roughness, which in the same way influences the critical current density. These measurements are in line with the results of Parkin *et al.* [7], where a similar lowering of the critical current density was measured for varying depinning fields from 15 Oe down to 5 Oe. The observed linear decrease of the current density with the depinning field is demonstrated for even less depinning fields.

Domain wall transformations that lead to an increased depinning field at current densities above the critical current density are discussed in the following. Periodic domain wall transformations and transformations into more complex domain walls are well known [70, 76] and may lead to a domain wall that requires a higher field to depin. For higher current densities, a second transformation may occur that brings the domain wall back to the original type (vortex wall), which can be displaced more easily. In our experiment for a bath temperature of 60 K, the first domain wall transformation occurs at current densities of more than 4.2×10^{11} Am^{-2} (inset of Fig. 5.1 (b), marked with an A). This can be deduced by the fact that larger current densities require larger depinning fields and also by

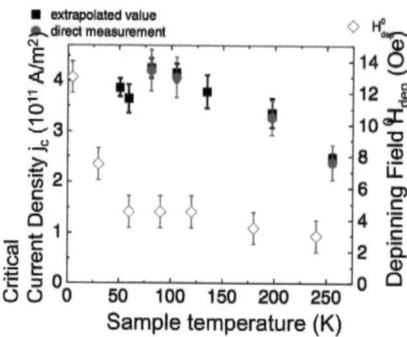

Figure 5.2: The critical current densities at zero field is the sample temperature are plotted by red circles for direct measurements. Extrapolated values, determined by linear fits from the data presented in Fig. 5.1 are plotted with black squares. Due to Joule heating, the minimum sample temperature of 50 K relates to a bath temperature of 4.4 K. The green open diamonds represent the depinning field in the absence of an applied current.

measuring the domain wall resistance, which is explained later and presented in Fig. 5.3. At current densities of about $\sim 7 \times 10^{11}\,\text{Am}^{-2}$, a transformation into a similar domain wall as that after the initialization is expected, as the critical field value reduces to zero again.

For a more detailed analysis of the real spin configuration, images of the domain state before and after the current injection would help to interpret the data. Such measurements for Py wires, fabricated with lift-off technique have been performed, where a clear dependence of the resulting domain wall transformation on the current pulse length and amplitude has been found [67].

In addition to imaging, the magnetoresistance (MR) measurements also give information about possible domain wall transformations and resulting domain wall states. When injecting currents at zero field, resistance changes are observed reproducibly. At a pulse amplitude of $5.4 \times 10^{11}\,\text{Am}^{-2}$, a resistance of $19.510\,\Omega$ is found and presented in Fig. 5.3 (up green triangles), which is lower than the initial $19.518\,\Omega$, after initializing the domain wall. In the case of a thick and narrow nanowire, vortex walls are the energetically favored state [58], so a transition into a transverse wall is not expected. Moreover, the reduction of the resistivity cannot be explained by the transition into a transverse wall [72], thus a modification of the vortex wall is a reasonable explanation. Further increasing of the current leads to a depinning at zero field at $6.6 \times 10^{11}\,\text{Am}^{-2}$. At even higher current densities,

5.1. Displacement in Materials with low Depinning-Field

the current leads to more deformation inside the domain wall and resistance values of 19.504 Ω are observed (down triangles). This value corresponds to half of the total AMR signal due to a single vortex. This points to a stronger change inside the spin structure or even a nucleation of a second vortex.

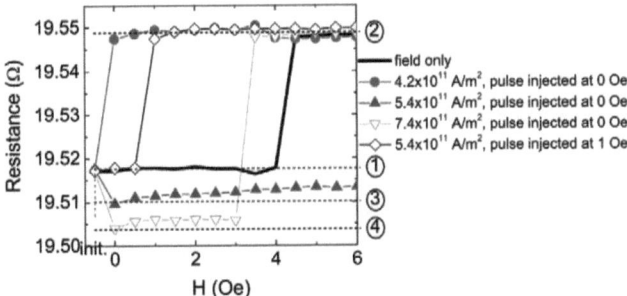

Figure 5.3: The resistance of a 900 nm wide and 25 nm thick Py wire as a function of applied magnetic field is plotted at a stable cryostat temperature of 60 K. The full black line corresponds to the field driven domain wall displacement without any injected current pulse. The red dots, green up triangles and red down triangles correspond to measurements where a single current pulse is injected at zero field with current densities of 4.2, 5.4 and $7.4 \times 10^{11}\,\text{Am}^{-2}$. A measurement with a single current pulse of $5.4 \times 10^{11}\,\text{Am}^{-2}$ applied at an applied field of 1 Oe, is plotted with blue open diamonds. The initial resistance is labeled with "init." and depicts the reproducibility of the domain wall structure after each initialization. The dashed lines, ①-④, represent different resistance levels, corresponding to the domain wall spin-configuration between the small Au contacts seen in Fig. 5.1. ① corresponds to the initial domain wall resistance (init.). ② corresponds to absence of a domain wall, ③ and ④ correspond to two different transformed domain wall spin structures. Note that the transformed domain walls have different depinning fields.

Complex spin structures, including possible nucleations and annihilations during the pulse injection can explain the change of resistivity and is described in similar systems in Ref. [71, 76]. The energy needed to depin a domain wall and the energy to nucleate a second vortex core and consequently to form a double vortex wall can be in the same range. The increased number of spins pointing perpendicular to the wire in the case of a multi vortex state can explain the change of the MR signal and is in line with micromagnetic simulations [77]. This explanation is corroborated by a measurement where a current pulse with a current density of $5.4 \times 10^{11}\,\text{Am}^{-2}$ is injected while a 1 Oe field is applied (open blue diamonds in Fig. 5.3). The small applied field tilts the spin configuration in such a way that the energy barrier for the depinning is lowered, whereas the required energy to nucleate a new vortex remains equal to the energy needed without an applied field. Now, in contrast to the measurement performed at zero field (green up triangles), the resistance increases to the value corresponding to no domain wall, indicating that the domain wall is depinned and moved out of the contacts.

5.1.3 Discussion

Permalloy wires with low edge roughness were patterned by Ar ion milling. The depinning field is found to be 2-3 Oe at room temperature and has a linear relationship to the critical current density at zero field.

The critical current densities observed are in the range of $2.5 \times 10^{11}\,\mathrm{Am^{-2}}$ at a cryostat temperature of 240 K and therefore 4-5 times smaller than in similar measurements performed with samples with the same geometry patterned with a lift-off technique. This very important result demonstrates the importance of the fabrication process used to pattern these nanowires, as extrinsic pinning effects have a strong influence on the critical current required for current-induced domain wall motion and the subsequent domain wall velocity. This is particularly important in device applications, where low current densities are required, which means that the pinning should be as low as possible.

Deformation of the domain wall spin configuration due to an applied current and the associated change of the depinning field were observed, as manifest in reproducible changes of the magnetoresistance. Due to this deformation, the domain wall cannot be easily displaced by a current above the critical current density. As a consequence, the current density needs to be accurately controlled when a reliable domain wall displacement is required.

5.2 Time-Resolved Domain Wall Motion

During the domain wall displacement, one can distinguish between the onset of the domain wall motion in the first few nanoseconds after the excitation, the actual domain wall motion, and the possible domain wall transformations. Domain wall transformations appear at very large current densities where also the destruction of the nanowire due to electromigration becomes likely. For a pump-probe experiment, which is needed to achieve high tempo-spatial resolution, a reproducible and reversible process is required. Therefore, the focus is set on the first response of the domain wall to an applied spin current. The first nanoseconds of a field driven domain wall displacement are presented in Chapter 4, where it was shown that inertia effects determine the onset of the domain wall motion, which is also expected for domain wall motion (similar to the delayed response in field-induced domain wall motion, see Chapter 4). The disadvantage of high applied currents is that it can lead to possible structural changes of the material. Also, the elevated temperatures can induce changes in the spin structure.

5.2.1 Experiment

The experimental setup is similar to the experiment performed in Chapter 4. However, instead of using a stripline to apply a magnetic field, the nanowire was contacted to the optical diode to apply current pulses. The diode is then triggered by a laser pulse. The pulse is synchronized with the orbit clock of the synchrotron and therefore a well defined delay of the current pulse to the camshaft bunch of the synchrotron can be set. The micro-channel plate (MCP) can be triggered by a 280 V pulse, to selectively amplify only the electrons that are generated on the sample surface during the camshaft bunch. XMCD images taken with this mode have a time resolution that only depends on the full width half maximum of the camshaft bunch, which is typically in the range of 50-70 ps.

A possible problem is pinning of the domain wall. A pinned domain wall, which is subjected to a current can be depinned by thermal excitation. If the magnetic field is not strong enough to trap the domain wall and the current density is too low for an assured displacement, very small thermally activated displacements in current direction can occur that, over minutes or even hours, displace the domain wall up to a few micrometers. To not misinterpret those data where creep motion simulates *in situ* domain wall motion, an image at a delay prior to the pulse was taken after each time resolved scan.

The sample is placed on a sample holder as shown in Fig. 5.4. This setup is tailor designed for *in situ* experiments. The chip is glued on a circuit board with UHV compatible silver glue and wire bonded to the Au contacts of the circuit board. A photodiode is connected to an electric contact (labeled in Fig. 5.4 (b)) of the PEEM holder and to the circuit board, and triggers a voltage when the laser pulse hits the diode (similar to the setup shown in Fig. 4.3 with the nanowire connected to the diode, and a coil below the sample instead of the permanent magnet).

The charge generated in the photodiode during the light pulse passes through

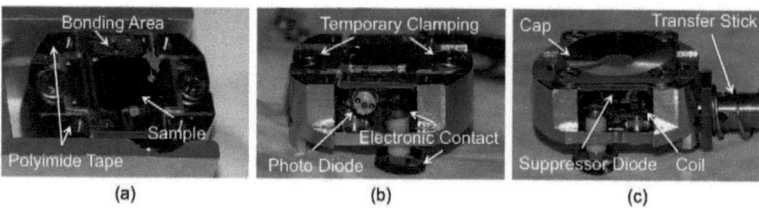

Figure 5.4: Sample holder for dynamic *in situ* experiments: In (a), a top view of the sample holder is shown. The sample is glued with silver glue on the specially designed UHV compatible circuit board. (b) shows a side view of the sample holder. This side faces up when the sample is mounted into the PEEM. The photo diode is placed below the circuit board and is soldered with UHV compatible tin to the board. The electronic contact to the diode and to one side of the coil is done by the spring on the foot. A ceramic feedthrough isolates this contact from ground, which is the second contact and directly connected to the sample. Two washers that clamp the circuit board are screwed on the holder to fix the sample during the wire-bonding. In the side view of (c), which is the side facing down when mounted into the PEEM, one can see the small suppressor diode, which is directly mounted to the board. The coil inside the construction can also be seen. The cap is now mounted and the structure is perfectly aligned in the center of its circular opening. Another polyimide piece is taped below the cap at the position where the bond wire connects the structure to the diode to ensure no shorts. The foot on this side is connected to the other side of the coil. Using the transfer stick, the sample is ready to be loaded into the PEEM load lock chamber.

the structure and induces the domain wall motion. The current generates a change of the local electric potential of the wire and the Au contact. This voltage is added to the start voltage of the PEEM, resulting in a change of the local intensity.

A delay scan, where the observation time is shifted over the current pulse is used to measure the shape of the current pulse. The total electron yield at a Au contact during the pulse varies due to the changing electric potential, which is a measure for the pulse shape. An intensity delay scan is presented in Fig. 5.5. The current at any time can be determined by the total current flow and the surface below this intensity delay scan.

A coil, placed below the sample is used to a tunable in-plane field during the experiment. This field is needed to create an attractive potential for the domain wall at the wire bend. With this configuration the domain wall displacement can be restored reproducibly. The field which is needed to trap the domain wall in the potential depends on every observed domain wall and needs therefore to be determined during the experiment. This is an improvement compared to the field-induced experiment in Chapter 4, where exchangeable permanent magnets were

5.2. Time-Resolved Domain Wall Motion

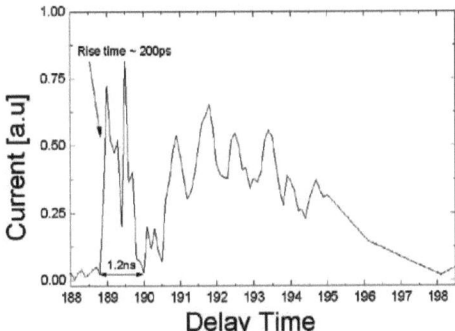

Figure 5.5: The total electron yield (PEEM) of the Au pad, which is connected to the Permalloy wire. The current pulse adds a potential, resulting in a change of the intensity. A delay scan over the current pulse yields information about pulse height and pulse shape. The graph shows a very steep rise time of the current pulse of around 200 ps. Due to reflections in the setup, a second intense pulse is generated that remains over 7 ns.

used, resulting in a time consuming exchange procedure. In the case of too high currents and the domain wall moving out of the wire, a large reversed field was applied to restore the domain wall.

PEEM imaging was chosen for this type of experiment, as the large ohmic losses need to be dissipated to a bulk substrate, which makes transmission techniques on membranes difficult. The disadvantage of PEEM is the presence of a high voltage of typically 20 keV between the sample and the PEEM. This voltage is applied between the sample and the objective lens of the microscope (see Section 2.1.1) and sometimes a discharge occurs, destroying the contacted structure. The photodiode is also at risk, but a suppressor diode usually disrupts the high current peak. To prevent these discharges, the voltage was reduced to 15 keV. The vacuum due to the complicated setup is usually worse than 5×10^{-9} mbar, which additionally enhances the probability of a discharge. In this experiment only one structure is connected at a time, and remounting or re-bonding the sample is required for measuring a different structure.

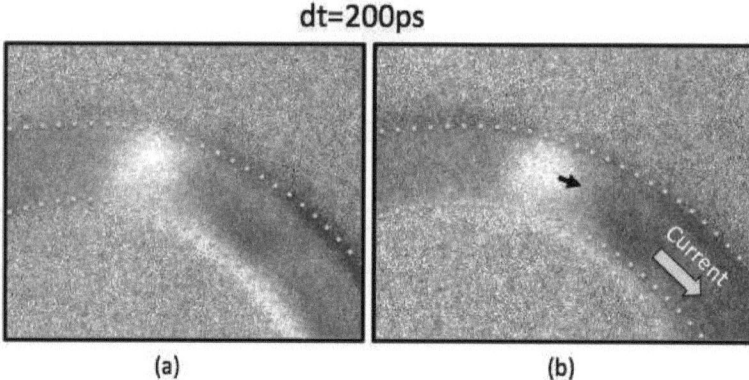

Figure 5.6: Time-resolved images of a domain wall shortly before the field pulse (a) and 200 ps later (b). The vortex core shifted slightly to the right, along the current direction. The wall did not depin at the used current density.

5.2.2 Results

The results of the time-resolved current-induced domain wall displacements are presented in this section. The problems of the varying start voltage and distortion due to the potential changes and possible generated Oersted fields make a precise analysis difficult. Most of the small displacements and distortions are only qualitatively analyzed. The domain wall does not depin at small current densities and short currents pulses. Therefore, the laser intensity and the gated voltage were increased. The increased laser intensity leads to more free charge carriers that are generated in the diode. The resulting current pulse is shown in Fig. 5.5. This intensity delay scan of the contacting Au pad shows a steep and short current pulse, that is followed by a longer and also intense current pulse, which can be most likely assigned to signal reflection in the setup due to impedance mismatch. A very high laser intensity results in the pulse shape shown (Fig. 5.5), and only at these intense pulses a domain wall displacement was achieved.

Before a domain wall can be imaged in a time-resolved manner, the critical current density for the *in situ* domain wall displacement, which is different for every single wire, needs to be determined. Therefore, the laser intensity is increased, until the domain wall displaces. The largest displacement in a field defined potential can be achieved when the applied magnetic field is just strong enough to keep the wall in the nanowire. In order to determine this value, a field is applied

5.2. Time-Resolved Domain Wall Motion

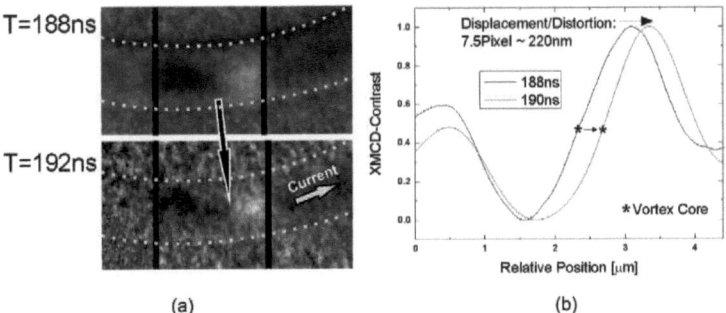

Figure 5.7: (a) Time-resolved XMCD images of a vortex wall. The electron current flows to the right and displaces the domain wall a few nm. The black bars indicate the outer border of the domain wall. The black arrow displays the displacement of the vortex core. The dashed yellow line is a guide for the eye that indicates the edge of the nanowire. A line scan over the domain wall at these two times is shown in (b).

together with a current pulse that is strong enough to depin the wall. Subsequent decreasing the field and imaging until the domain wall starts to depin reveals the smallest field required to force the domain wall back to the equilibrium position. After that, the domain wall was initialized again and a field slightly above this threshold was set.

In Fig. 5.6, a domain wall is shown which was modified by the current pulse. The vortex core moved a few nm to the right, although the absolute position of the domain wall seems to be unchanged, meaning that only the spin structure of the wall is slightly distorted. The time-resolved XMCD image of another domain wall is shown in Fig. 5.7. Here, the delayed image is taken 4 ns after the first pulse starts (a). This delay time is already on the second pulse (see Fig. 5.5), as the total displacement at earlier times was too small to image. The domain wall displacement is now clearly visible and the domain wall is shifted slightly in the current direction, to the right. The XMCD contrast intensity line scans in (b) show that the vortex core and the right side of the domain wall are displaced by ~ 220 nm, which corresponds to a domain wall velocity of 55 m/s, a value which is in line with other experiments [25].

5.2.3 Discussion

Short and intense current pulses are used to excite domain walls in a Permalloy nanowires. Time-resolved PEEM imaging was used to map the spin configuration during the pulse, and domain wall velocities up to 55 m/s on short time scales were observed, in line with similar experiments [25]. The reduced effective exposure time due to the time-resolved mode and the distortions, originated by the high applied voltage and current, make PEEM imaging challenging. Destructive discharges, requiring for re-bonding the sample, make the experiment even more difficult.

In conclusion, *in situ* time-resolved imaging of domain wall motion with XMCD PEEM is possible but challenging. The distortion of the images cannot always be corrected, with a highly decreased magnetic contrast, which along makes a quantitative analysis problematic. Possible improvements for this type of experiment would be the use of a material composition that shows smaller critical current densities. This would also allow the application of STXM, which does not uses high voltages. Up to now, the Joule heating of the intense current pulses is dissipated by the bulk substrate of a Si chip, which is not as efficiently dissipated in the case of a thin STXM membrane.

5.3 Conclusion

Current induced domain wall motion in Permalloy nanowires is a widely studied topic [25, 65–68]. For applications based on domain wall motion, reproducible displacements and low critical current densities are a prerequisite.

The critical current density could be reduced by excluding domain wall pinning. A specialized patterning method was used to fabricate smooth wires with less edge induced pinning that showed a reduction of the critical current density by a factor of four. However, for practical applications, a reproducible displacement at a well defined domain wall state is crucial. At current densities above the critical current density, domain wall deformations have been observed. These domain wall deformations change the critical current density for each domain wall and can lead to the annihilation of domain walls.

The reduction of the critical current density by an alternative patterning method demonstrates the importance of the fabrication process. Further improvements in the magnetic material and in the patterning method may suppress

5.3. Conclusion

further the critical current density to finally reach a value that allows for current induced domain wall motion in device applications.

Domain wall motion in ordinary lift-off patterned nanowires was imaged in a time resolved manner to map the spin configuration of a domain wall at the early stage of the domain wall, driven by a spin current. The used currents were too low to displace the domain over a large distance, and the experiment was limited to small excitations and wall displacements. The images, taken during a time scan, show a very fast domain wall motion in the range of 55 m/s.

CHAPTER 6

Spin Configuration in Patterned LSMO

Half-metals are characterized by one semiconducting or insulating and one metallic spin band, rendering the conduction electrons fully spin-polarized at the Fermi level. This property makes this class of materials ideally suited for applications such as emitters of spin-polarized electrons for spintronics devices and magnetic random access memory elements based on the tunnel magnetoresistance effect [10]. The optimally doped $La_{0.7}Sr_{0.3}MnO_3$ (LSMO) is a *transport half-metal* with itinerant ↑ electrons and localized heavy ↓ electrons at the Fermi energy [78]. It is characterized by a measured spin polarization of nearly 100% [79] and a Curie temperature T_c up to 370 K [80], making this material a good candidate for room temperature device applications [81, 82].

LSMO belongs to a class of compounds known as the "colossal" magnetoresistance (CMR) manganites, which are characterized by a rich electronic behavior as a function of chemical doping, strain, as well as external magnetic and electric fields. This sensitivity to external parameters makes the CMR manganites particularly suited for studying a wide range of physical phenomena, such as the role of charge carrier density on the spin configuration in multiferroic heterostructures [83, 84], the metal to insulator transition [85], and the role of strain on the equilibrium magnetic state [86].

The study of static and dynamic properties of domain walls in this class of materials is also of interest for applications in mass storage concepts based on domain walls [6–8], since the high spin polarization P together with low saturation magnetization M_s at temperatures near the Curie temperature T_c suggest a high spin-torque efficiency for current-induced domain wall motion (which scales as P/M_s) [21].

For the realization of experiments and applications based on single domains or domain walls, one needs a suitable system with controllable domain states. This is best accomplished in systems with low magneto-crystalline anisotropy, where the spin configuration can be controlled by the shape anisotropy. However, several high spin polarization materials investigated show spin structures that are strongly dominated by magnetocrystalline anisotropy, which is unsuitable for many devices that rely on domain walls [87, 88].

6.1 Sample Description and Characterization

LSMO thin films can be grown epitaxially on either bulk $SrTiO_3(001)$ (STO) or STO buffered Si(001) substrates. Two different approaches to pattern these epitaxial LSMO thin films are developed, one with focussed ion beam lithography and one with EBL defined Cr seeded substrates. In this section, the two different sample types are introduced and the characterization is presented.

The LSMO films were deposited by pulsed laser deposition from a stoichiometric target onto $SrTiO_3(001)$(STO) single crystal substrates by L. Méchin and F. Gaucher in Caen, France. The laser radiation energy of the KrF excimer laser (248 nm) was 220 mJ at a repetition rate of 3 Hz. The oxygen pressure was 0.35 mbar and the substrate temperature was held at 720°C during growth. These parameter values were found to be optimal for producing single crystal films with a smooth surface, as judged by X-ray diffraction (XRD) and atomic force microscopy (AFM). The XRD data shows highly [001]-oriented LSMO films. SQUID magnetometry shows that the magnetization is in-plane, with a Curie temperature of about 340 K (see Fig. 6.1), typical for good quality films of this composition.

The quality of an LSMO film can be estimated by the comparison of resistivity measurements with field cooling measurements of the magnetization (SQUID). Here, the temperature dependent resistivity shows a positive slope, characteristic for a metallic system. It shows an inflection point near T_c, which marks the transition into an insulating regime at higher temperatures.

6.1. Sample Description and Characterization

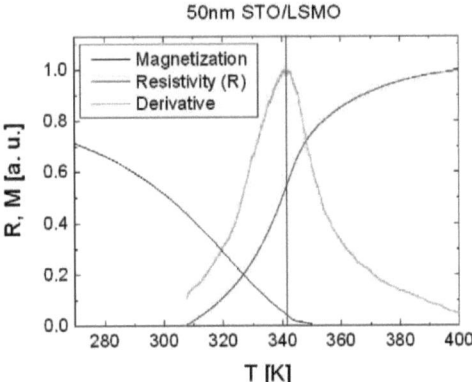

Figure 6.1: Resistivity R as a function of temperature (red) and magnetization M (black, SQUID) of a 50 nm thick LSMO film. The derivative of the R(T) curve is plotted (green). The blue line marks the temperature corresponding to T_c as measured by SQUID and the highest slope of the R(T) curve.

Fig. 6.1 shows the comparison of these curves for a 50 nm thick LSMO film. The saturation magnetization of the samples decreases to zero at the Curie temperature T_c (black curve). At the same time, the resistance of the film (red curve) increases. The derivative of the resistivity curve, shown in green, has a maximum at around 341 K, marking the highest slope of the resistivity curve. The blue line marks this temperature and one can see the good agreement with the T_c measured with SQUID. LSMO films with these characteristics are of very high quality due to the ideal epitaxial growth conditions.

The design patterned on the films, was chosen to allow the investigation of the affect of magnetocrystalline anisotropy on the magnetic state and to investigate the domain and domain wall configurations in LSMO. The patterns consisted of basic shapes (triangles, squares, rings and rectangles), which were varied in size and were also rotated by 45° and 90° to investigate the effect of changing the orientation of the anisotropy axes. The two patterning methods used that were introduced in Section 3.2.5 are recalled next:

FIB patterned LSMO

LSMO films, with thicknesses varying from 15-100 nm, are patterned using FIB-lithography (see Section 3.1.3). By looking at the SEM contrast during the FIB lithography, an indication was obtained about which doses were required to remove the material. The doses used to pattern the first batch of LSMO films

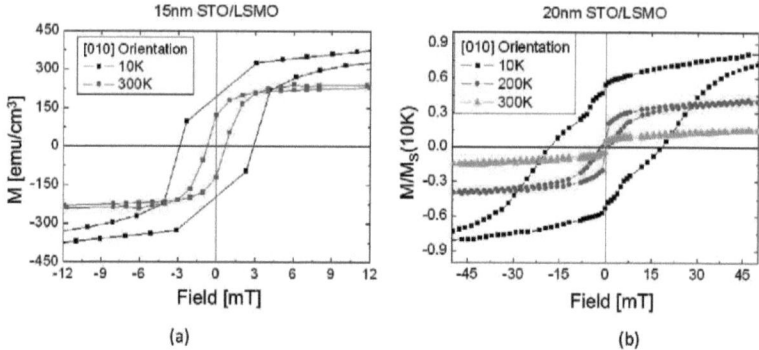

Figure 6.2: (a) Magnetic hysteresis loops, for a 15 nm LSMO thin film along the [010] direction at 10 and 300 K measured with a SQUID magnetometer. The saturation magnetization increases by a factor of two, when cooling down to 10 K, while the coercive field increases by about a factor of five. (b) hysteresis loops for a patterned 20 nm LSMO film along the [010] direction at 10 K, 200 K and 300 K. The coercive field increases from 0.4 mT at 300 K to 18 mT at 10 K. The saturation magnetization also increases by a factor of four, upon cooling. The pronounced steps in the loops correspond to different coercive fields from different patterned areas.

was therefore chosen to be in a range of 1000-3000 μCcm^{-2}, depending on the film thickness. First experiments have shown that the removal of material is not necessary since much smaller doses are sufficient to transform the ferromagnetic LSMO into the paramagnetic state. A second batch was successfully patterned using this material damage method (rather than removal) with doses in the range of 50-100 μCcm^{-2}.

EBL patterned LSMO

LSMO needs ideal conditions to grow epitaxially on STO. In particular, LSMO requires an appropriate substrate with a small lattice mismatch to be grown as a single crystal, otherwise LSMO films will grow amorphous or polycrystalline. To define single crystalline structured elements in a non-epitaxial matrix, a Cr seed mask is patterned via EBL (see section 3.2.5). Cr is chosen as a patterned seed layer to induce amorphous growth of the LSMO so that only the LSMO on the uncovered STO substrate grows epitaxially. LSMO films varying in thickness from 20-60 nm in 10 nm steps were subsequently grown on these pre-patterned substrates. The EBL patterning only seemed to work as intended for the 20 nm thick sample. This sample showed a large contrast in the optical microscope, while the other thicker samples exhibit a much weaker optical contrast. One likely cause

6.1. Sample Description and Characterization

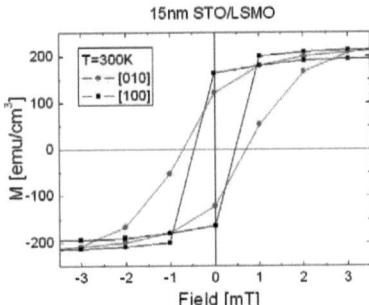

Figure 6.3: Hysteresis loops of a continuous 15 nm thin film along the [100] and [010] directions, measured by SQUID magnetometry. The variation of the coercive field depicts the uniaxial anisotropy, which is along the [100] direction.

is that, for the thicker samples, the LSMO growth on the Cr is polycrystalline rather than amorphous. It is only the amorphous material in the 20 nm thick sample that leads to a significant contrast difference.

Characterization

SQUID measurements are performed in order to characterize the magnetic properties of the material. In Fig. 6.2, hysteresis loops, measured by SQUID magnetometry are presented. In Fig. 6.2 (a), hysteresis loops of a 15 nm thick film are shown, measured at 10 K (black curve) and at 300 K (red curve). Both the coercive field and the saturation magnetization increase at lower temperatures, typical for a ferromagnetic system. In Fig. 6.2 (b), hysteresis loops of the 20 nm-thick EBL patterned sample are shown. In addition to the increasing coercive field and saturation magnetization at lower temperatures, several pronounced steps are visible in the loops. This sample is the only one of this series that shows amorphous LSMO growth on the chromium seed layer. This explains the steps, which are induced by the magnetization of the patterned elements that have different switching fields, due to the different shape and size.

To see the influence of the magnetocrystalline anisotropy, SQUID magnetometry measurements on the 15 nm-thick film were performed along [100] and [010] directions as shown in Fig. 6.3. The measurement along the [010] direction is plotted in red and shows a slightly larger coercive field, than the [100] direction (black curve). The sharp switching along [100] versus [010] indicates that the [100] axis is the easy axis. It is expected that LSMO has a fourfold magnetocrystalline anisotropy. However, one direction dominates and the magnetic effects can be

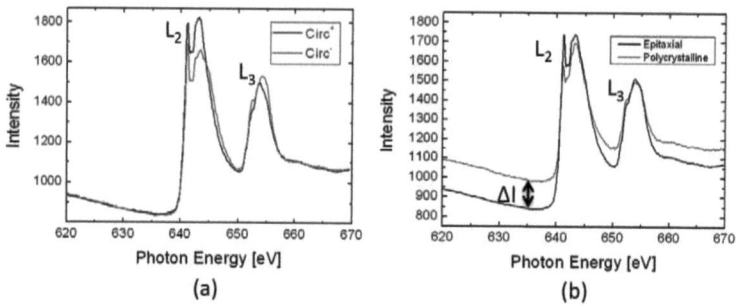

Figure 6.4: (a) The X-ray absorption spectra (XAS) of a uniformly magnetized area is plotted for both light polarizations as a function of photon energy. (b) The XAS for the epitaxial area is compared with the XAS of the polycrystalline area. A huge difference in intensity at energies below the L_2-edge is used to distinguish the structured area from the polycrystalline background, which shows the same contrast at the L_2-edge (see Fig. 6.5 (a) and (b)).

described by assuming a uniaxial magnetocrystalline anisotropy [45].

6.2 Magnetic Imaging of LSMO with PEEM

The magnetic imaging of the LSMO samples was conducted with PEEM.[1] Charging was avoided by electrically contacting the sample surface to the sample holder. Only the 20 nm LSMO sample, prepared with a Cr-mask showed charging effects, and therefore required the deposition of a conductive Cu capping layer. In total, 3 nm Cu was put on the surface to eliminate this problem.

A weak topological contrast of the Cr pre-patterned LSMO samples is present when imaging with X-rays at the manganese-L_2-edge. In Fig. 6.4 (a), the two X-ray absorption spectra are plotted for circular polarizations, showing a dichroic splitting at the L_2 and the L_3-edges. Contrast values of up to 4.5% are observed at the L_2-edge.

At energies below the L_2-edge, a big difference in absorption exists between the polycrystalline and the epitaxial area, which is shown by the absorption spectra in Fig. 6.4 (b). Here, the total electron yield of the polycrystalline area is compared with the electron yield of the epitaxially grown film. This change of intensity can be attributed to the conductivity difference between the two regions, which

---
[1] SLS (Villigen, CH), Bessy (Berlin, D) and Elettra (Trieste, I)

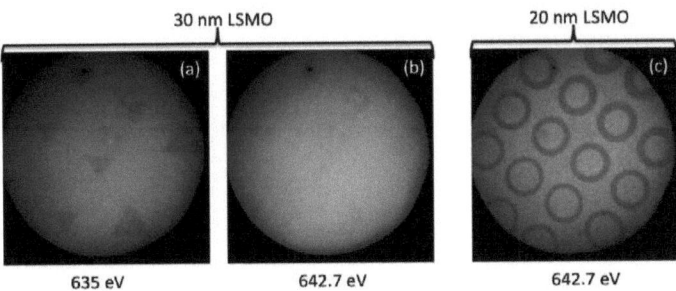

Figure 6.5: X-ray absorption image of a patterned 30 nm thick EBL patterned LSMO film, imaged at a photon energy below the absorption edge at (a) 635 eV and at the absorption edge, (b) 642.7 eV. The weak remaining contrast in (b) originates from the high XMCD-contrast observed in the film. The large difference in the absorption at energies below to the edge is used for finding structures and to optimize the alignment of the PEEM, which is demanding for a clean surface that is nearly free of topological contrast. For the 20 nm sample, a much higher contrast at the L_2-edge was found (c).

causes a change in the potential and therefore of the start voltage that influences the total electron yield [89]. By tuning the insertion device to a lower energy, the contrast of the elements (compared to the polycrystalline area) is increased. This effect is used to find the elements and to fine tune the focus of the PEEM image. In Fig. 6.5 (a) and (b) this drastic effect of increasing contrast is shown for a 30 nm thick LSMO film for an energy below the L_2 peak at 635 eV and at the absorption peak at 642.7 eV. In the case of the 20 nm LSMO film, a higher contrast in Fig. 6.5 (c) is observed at the L_2-edge, due to the amorphous growth conditions on the Cr-seed layer.

6.3 Domain Configurations

The evolution of the magnetic states of LSMO elements were studied as a function of size, shape, thickness, and temperature, showing that for sufficiently small elements, the equilibrium states consist of flux closure states, which are stable up to T_c. In addition, well defined domain walls were found in confined structures, demonstrating that the spin structure can be tailored by the element geometry, making it an ideal candidate for spintronic applications, such as spin emitters or spin-torque based devices.

Depending on the investigated material, different possible spin configurations are observed in patterned thin films. The dimensions of the elements play a big role, in addition to the presence of an applied field or the sample temperature. This means that the history of the sample, i.e. whether it has been magnetized or heated, is important for the resulting spin structure. So, for example, in one single magnetic ring, many different spin configurations can be found at one temperature and at remanence, after treating the sample in different ways.

A simple phase diagram, for example, that predicts the existence of either transverse or vortex walls inside a nanowire, depending on the geometry [58, 90], is not universally valid if the sample is magnetized or heated in different ways. At elevated temperatures (T $\gg$ 0 K) thermal effects play a big role for the spin structure, which is an important issue in modern data storage devices. At elevated temperatures, transformations of domain walls and even completely new domain patterns may occur [90].

The LSMO system is ideally suited to study these various effects. The low T_c ($\sim 60°C$) can be easily reached in a short time *in situ* in the PEEM. The sample drift is well controlled at these moderate temperatures, and consequently, imaging during the heating or cooling is possible. Not only the T_c is not too high, but also the coercive field H_c and depinning field H_{dep} can be easily reached with small magnetic fields produced by a coil on the sample holder inside the PEEM. An image of such a holder is presented in Fig. 5.4 in Section 5.2.1.

With the small coercive field, low Curie temperature and the weak uniaxial anisotropy in the LSMO films, different effects can be studied. The structured pattern was designed to depict magnetocrystalline anisotropy effects. Therefore, basic shapes are patterned in different orientations with respect to the crystalline axis. Rings and nanowires are patterned in order to allow one to image and to manipulate domain walls with both field and temperature.

6.3.1 Domains in Square and Triangular Elements

The SQUID data in Section 6.3 indicates that the thin LSMO films have a uniaxial anisotropy. To investigate the influence of this uniaxial anisotropy, different shapes are investigated in detail to find possible breaking of the magnetic configuration symmetry and effects on the spin structure that are induced by the uniaxial anisotropy. Triangles that have an edge parallel to the anisotropy axis are expected to have a distorted Landau state and squares that are rotated and unrotated in respect to the anisotropy axis are likely to have different domain states.

Domain States in Square Elements

The role of the uniaxial anisotropy can be seen directly in squares with varying size. The XMCD-PEEM images on the left of Fig. 6.6 are compared with a micromagnetic OOMMF simulation on the right, with a uniaxial anisotropy contribution with easy axis along the [100]-direction. In the experiment (Fig. 6.6 (a)-(d)), as in the simulation shown in Fig. 6.6 (e), the spins tend to align parallel to the easy axis. The difference between the experiment and the simulation is that in the case of the experiment, the vortex core is off-center in the larger elements (Fig. 6.6 (b) and (c)), and even in the smallest element (Fig. 6.6 (a)), a slight dominance of the black domain can be observed.

For bigger elements (Fig. 6.6 (d)), even more complex flux closure domain pattern, such as the diamond state, are observed. Interestingly, most of the generated 180°-domain walls appear to be black in all observed elements taken for this sample during this measurement. The most reasonable explanation is that the sample has been randomly exposed to a small magnetic field when the domain pattern has formed. The reason for the fact that the vortex core remains displaced is not trivial, because micromagnetic simulations shows a relaxation of the vortex core back to the center when the field is switched off (see Fig. 6.6 (e)). The easiest explanation for this behavior might be pinning. However, the fact that these 180° domain walls appear regularly can be explained by the energy difference of a vortex core at the end (Bloch point) and in the center of such a wall.

Theoretically, starting a numerical simulation of a square element with the spins oriented in a Landau state configuration, a centered vortex core will remain in the system, even for systems with cubic or uniaxial anisotropy. This counts for every orientation of the element with respect to the anisotropy axis. However, this was not observed in the flux closure states of the elements that have parallel edges to the magnetocrystalline anisotropy (Fig. 6.6). In the elements with a 45° angle in respect to the magnetocrystalline anisotropy a similar offset of the vortex core is observed in many elements. One example for a non-centered vortex core in a rotated element is shown in Fig. 6.7 (a). In general, the domain configuration for rotated elements (Fig. 6.7) is very different from those of the unrotated elements (Fig. 6.6). The domain states, presented in Fig. 6.7 are remanent states after the sample was cooled down from a temperature above T_c. In order to understand

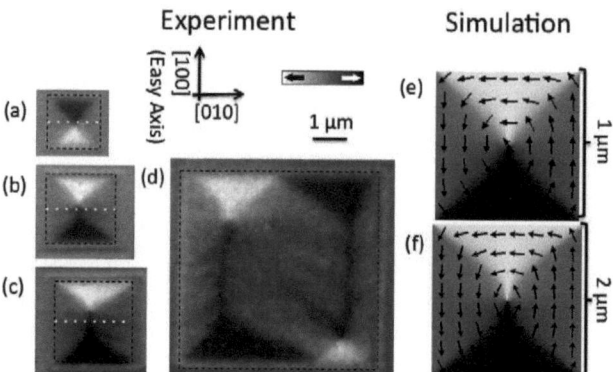

Figure 6.6: Domain states in 15 nm FIB patterned LSMO. A weak uniaxial anisotropy with an easy [100] direction influences the spin configuration. A numerical simulation is carried out to compare with the experimental results.

the observed domain states OOMMF simulations are performed by relaxing the system starting from a random magnetization until a spin configuration was found that represents the experimental findings [17]. Three complex domain states that were reproduced by these micromagnetic simulations are presented in Fig. 6.7. For these simulations, the saturation magnetization M_s was 200 kA/m, the exchange constant A was 2.7×10^{-12} J/m^3 and the uniaxial anisotropy was set to K=0.4 kJ/m^3.

Although the spin configuration is not absolutely identical to the experiment, the simulations show that starting from a random configuration the generation of these complex but stable spin configurations is achieved. The comparison of the experiment with the simulation reveals very subtle details. In Fig. 6.7 (a), a non-centered vortex core in a 2.2 μm large square element is visible in a flux closure state. The randomly relaxed state in Fig. 6.7 (b) shows a very similar state. Here, we see a small second vortex in the right corner with a chirality opposite to that of the big vortex at the center. The domain wall contrast direction in this image appears black and white, whereas the domains itself appears mainly gray. The direction of the chirality is visible in the order of the black and white domain walls. A close look at the experimentally observed domain state in Fig. 6.7 (a) shows the same small vortex in the right corner. Both vortices, in the experiment and in the simulation are marked with a black circle.

In Fig. 6.7 (c), a 3.3 μm large element is shown and compared with a simulation

6.3. Domain Configurations

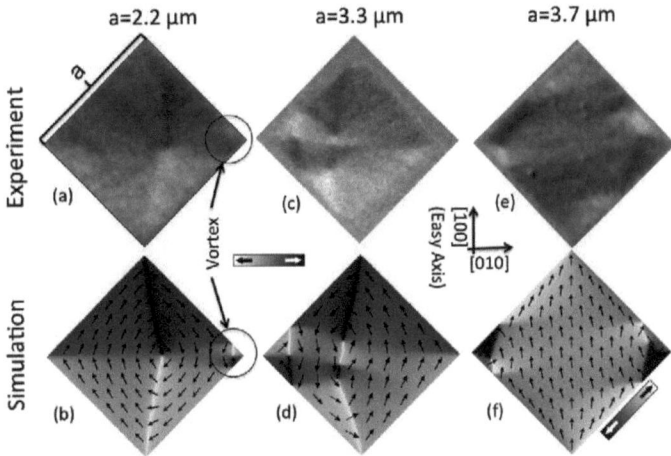

Figure 6.7: Examples of XMCD-PEEM images of complex domain structures in LSMO compared with numerical simulations. (a) shows an XMCD image and (b) the numerical simulation of a Landau-like state with one big and one small vortex. (c) XMCD image of a triple vortex state, which is in good agreement to the simulation (d). An XMCD image of a rather large structure in (e) shows two vortices, placed in diagonal opposite corners of the element, which was reproduced by the simulation in (f).

in Fig. 6.7 (d). This stable triple vortex state is more complex than the double vortex state, demonstrating the reduced influence of the shape anisotropy.

An XMCD image of a 3.7 µm element is shown in Fig. 6.7 (e). Two small vortices are placed in two diagonal opposite corners. The simulation of a large square element in Fig. 6.7 (f) demonstrates a similar behavior. The easy axis forces the spins to be aligned in the [100] axis and only at the two left and right corners the stray field energy can be reduced by the formation of two vortices.

The existence of a flux closure spin structure is primarily determined by the dimension of the element and the material parameters, but also on the field and thermal history of the element. In Fig. 6.8, examples of domain states in 15 and 50 nm FIB patterned LSMO squares and rectangles with varying sizes are shown. The sample was magnetized with ∼0.1 T along the easy [100]-direction prior to the imaging. In Fig. 6.8 (a), the domain states of the 15 nm thick film show flux closure domain states up to dimensions of 2 µm. The 50 nm thick domains in Fig. 6.8 (b) instead, show flux closure domains up to dimensions of 4 µm. For larger element sizes, when the shape anisotropy becomes less relevant, the domain pattern becomes more complex and C-states are observed regularly. In Fig. 6.8 (a), the spin direction of a Landau- and a C-state is marked with white arrows. For very small elements, single domain states are also observed. One example is shown in Fig. 6.8 (a) for a 300×500 nm rectangle.

This thickness dependence of the stability of flux closure states can be verified by micromagnetic simulations. An element with initial Landau spin configuration is exposed to a magnetic field. A vortex annihilation field can be defined, which is the field required to annihilate the vortex core and therefore to change the remanent domain state to a non flux closure state. Simulations of this field as a function of edge length and film thickness are shown in Fig. 6.9.

OOMMF simulations of the thickness dependent vortex annihilation field is presented for 5 nm to 100 nm thick films in Fig. 6.9 (a) [17]. The vortex annihilation field increases linearly with thickness. Figure 6.9 (b) shows micromagnetic OOMMF results of the field dependence for a 15 nm and a 50 nm film for element sizes from 0.25 µm to 3 µm. The saturation magnetization M_s was chosen to be 200 kA/m, the exchange A was 3×10^{-12} J/m^3 and a uniaxial anisotropy with K=0.1 kJ/m^3. A decreasing slope is observed, which is fitted with a first order exponential decay. The trend of these curves agree with the experimental findings on the 15 nm and 50 nm LSMO films, which show C-states after magnetizing above a certain threshold size, which is larger for the 50 nm than for the 15 nm

6.3. Domain Configurations

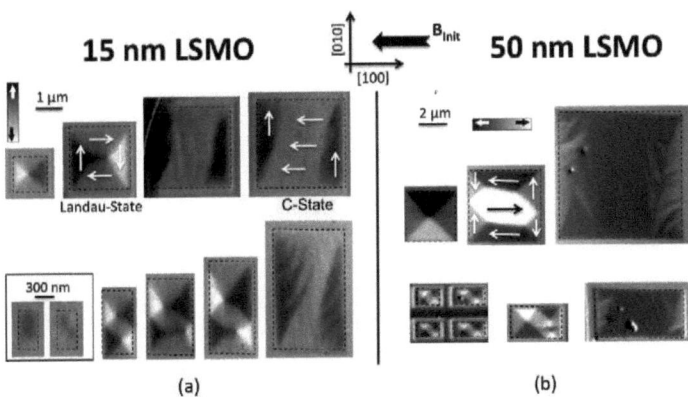

Figure 6.8: Spin configurations in two different FIB patterned LSMO films are shown after the sample was magnetized with ~0.1 T. Some representative states are labeled with arrows, to indicate the spin orientation. In (a), square and rectangular elements with different dimensions on a 15 nm thick film are compared. The flux closure spin structures are stable up to element dimensions of $2\,\mu$m. The left-most rectangle has a short edge of 300 nm and is in a monodomain state. (b) square and rectangular elements in a 50 nm thick film. The flux closure states remain stable up to typical dimensions of $4\,\mu$m.

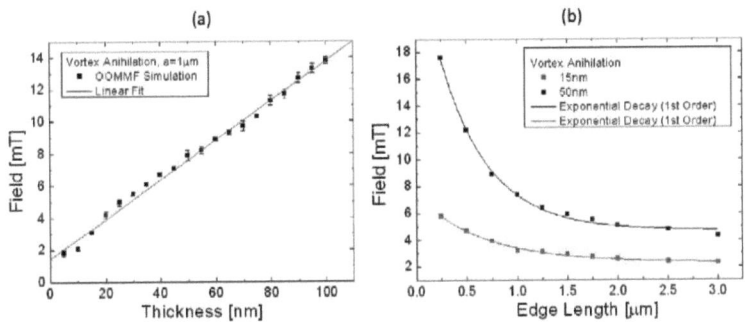

Figure 6.9: Micromagnetic simulations of squared elements: In (a) the vortex annihilation field is plotted for a $1\,\mu$m $\times\,1\,\mu$m square element. The dependence of the vortex annihilation field with the element size is shown in (b) for a film thickness of 15 nm (blue) and 50 nm (black). Both curves are fitted (solid line) assuming an exponential decay.

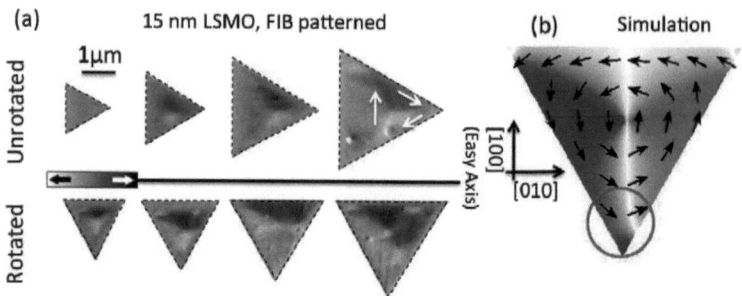

Figure 6.10: (a) XMCD image of unrotated and 90° rotated triangles. The uniaxial anisotropy points in [100] direction, therefore the vortices are forced in the direction of the edge parallel to the [010] for the triangles on top. The 90° rotated triangles have no edge parallel to the easy axis and show therefore more complicated domain structures. (b) Micromagnetic simulation of a 2 μm large rotated triangular element with a small uniaxial anisotropy.

thick film.

Domain States in Triangular Elements

In triangular elements, at least one of the three edges is aligned away from the easy axis. The spin configuration of a flux closure Landau state in an equilateral triangle is now dependent on the orientation of the element with respect to the crystal axis. One can differentiate between two cases: one where the easy axis is aligned to an edge and one, where no edge is aligned with the easy axis.

In order to observe effects that are induced by the uniaxial anisotropy and the respective element orientation, triangles with varying edge length from 1.6 μm to 3.5 μm were FIB patterned on the 15 nm thick LSMO film. For the non-rotated triangles, XMCD images are presented in the top of Fig. 6.10 (a). The smallest triangle is nonmagnetic, showing the influence of ion induced damage. The larger elements have a nonmagnetic frame due to ion irradiation. The domains aligned parallel to the easy axis ([100]-direction) are dominant, in contrast to the edges that have an angle of 60° with respect to this axis. This pushes the vortex core to the right side.

The bottom of Fig. 6.10 (a) presents triangles that are rotated by 90° in order to have one edge aligned perpendicular to the uniaxial anisotropy. One can see that the vortex cores are pushed towards the edge parallel to the [010]-axis, which

6.3. Domain Configurations

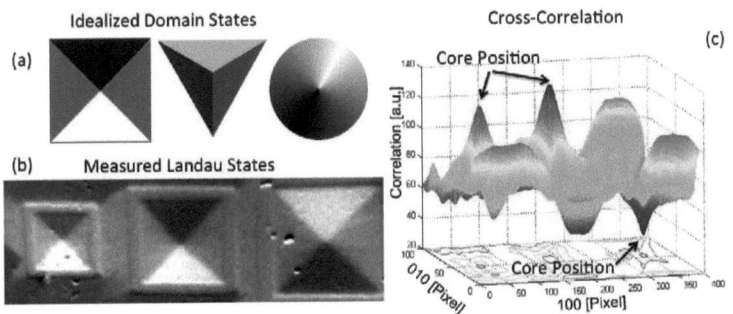

Figure 6.11: (a) Idealized domain States as comparison mask. (b) Example of a Landau state, measured with XMCD-PEEM. (c) Cross-correlation of an idealized Landau state with the example image shown in (b). The cross-correlation shows two maxima (red) at the vortices with the same chirality as the idealized vortex core. The minima (blue) indicates the vortex core of the Landau state that has an inverted chirality, compared to the idealized state.

has no component parallel to the anisotropy. Also an asymmetry can be observed, although the edges on the left and on the right have the same orientation with respect to the anisotropy axis. This behavior was also shown in the simulations in Fig. 6.10 (b). Here the spins at the bottom corner are non-symmetrically orientated, thus causing the asymmetry.

To quantify these results, the position of the vortex core is determined with a cross-correlation function, where the similarity of a spin configuration to an idealized image is compared and quantified. An example of a cross-correlation is shown in Fig. 6.11. A specialized Matlab program is used to compare any given shape with a spin configuration to find the relative position in an XMCD image. Idealized standard spin configurations are presented in Fig. 6.11 (a). For the three squares shown in Fig. 6.11 (b), the left Landau structure of Fig. 6.11 (a) was used. The corresponding cross-correlation in Fig. 6.11 (c) shows two maxima (red peak, marked with a black arrow) and one minima (blue, marked with a black arrow). A maximum is a vortex core of a Landau state that has the same chirality as the idealized shape. The minimum is a vortex core with opposite chirality.

This cross-correlation is now used to determine the vortex core position of the experimentally observed displaced vortex cores in the 90° rotated triangular elements. The determined position is normalized to the edge length a and plotted in Fig. 6.12 (a). The triangle is visualized with black lines.

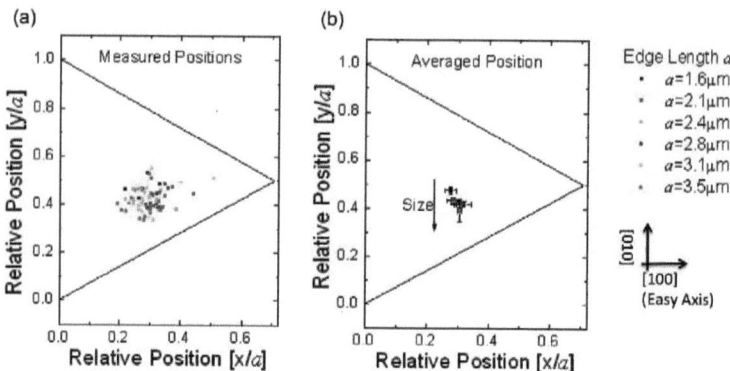

Figure 6.12: (a) Relative positions of the measured vortex core position of rotated triangular elements in 15 nm LSMO. (b) The average positions of the vortex core position is plotted. A size dependent trend of the relative displacement is indicated with a black arrow.

The average positions are plotted in Fig 6.12 (b). A trend of the relative vortex core displacement can be observed as a function of size (marked by the black arrow). This dependence can be explained by the contribution of the shape anisotropy, which becomes less important for large structures.

6.3.2 Domain Walls in Wires and Rings

Domain walls in micro- and nanowires are of scientific and technological interest. Depinning fields and critical current densities needed to depin and displace a domain wall depend both on pinning and on the particular domain wall type present in the wire [58]. The domain wall type depends on the material type, the geometry and also on the temperature. A detailed knowledge about domain wall configurations is crucial for further experiments and device applications.

For in-plane magnetized materials, ring elements with varying diameter, width and thickness are ideally suited for the study of domain walls. In the following sections, results will be presented on ring elements and zig-zag wires on both FIB and EBL structured samples.

In Fig. 6.13, rings prepared with the different techniques are shown. In (a), rings patterned with FIB are in an onion state that show vortex and multi-vortex domain walls. In (b) an onion state is shown that presents both a vortex and

6.3. Domain Configurations

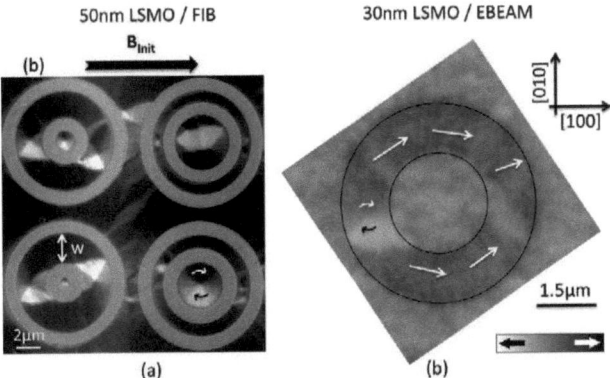

Figure 6.13: Spin configuration of onion states after magnetizing the sample with a large in plane field. In (a), FIB-patterned rings with vortex domain walls are shown. In (b), a 30 nm thick EBL defined element (by Cr prepatterning) is shown. On the left side a vortex- and on the right side a transverse domain wall is present.

a transverse domain wall. In order to generate such an onion state, the sample is magnetized in the plane in order to saturate the spins parallel to the field. By decreasing the field down to zero, the spins follow the shape of the ring and consequently form two domain walls, one head-to-head and one tail-to-tail. This domain state is called "onion" state [91].

By imaging a domain state under two different angles, the full in-plane spin configuration can be distinguished. In Fig. 6.14 (c), the full in-plane contrast is calculated from the two XMCD images in Fig. 6.14 (a) and Fig. 6.14 (b). The two very complex domain walls in the upper right and lower left, that define the onion state, are strongly influenced by pinning and also by the uniaxial anisotropy. Although a clear distinction between transverse and vortex wall is difficult, one can see a vortex in the upper domain wall, whereas the lower domain can be identified as a transverse wall.

In experiments that aim at studying the displacement of domain walls, rings are rather unsuited as the displacement is limited to the size of the ring and the two walls of the onion state can interfere. A more suited geometry, which is also particular interesting for applications or field and current induced experiments, are magnetic nanowires. Figure 6.15 presents domain walls in nanowires in a 30 nm thick EBL patterned LSMO film. The sample is magnetized in the direction perpendicular to the bend of the wire. Domain walls can be reproducibly

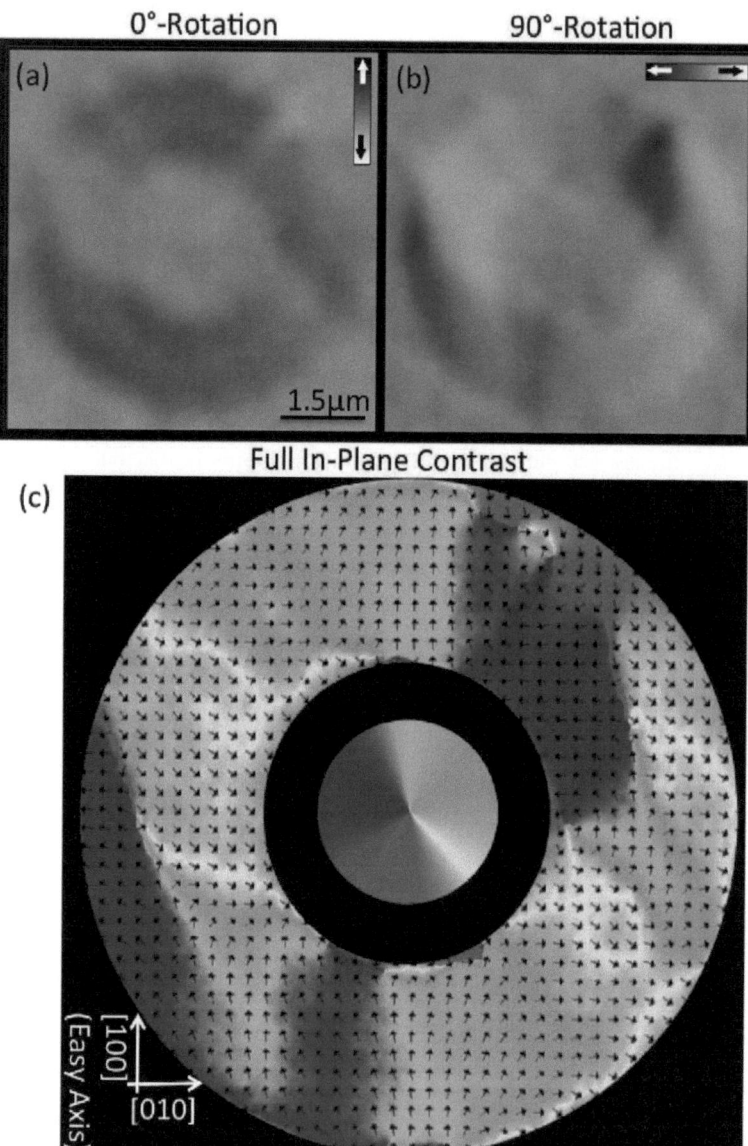

Figure 6.14: The XMCD images of a 30 nm thick and 1.5 μm wide LSMO ring element, taken under two perpendicular directions, are shown in (a) and (b). The full in-plane contrast, extracted from these images is presented in (c). The color-code for the spin orientation is given in the center. The spin direction is indicated by black arrows.

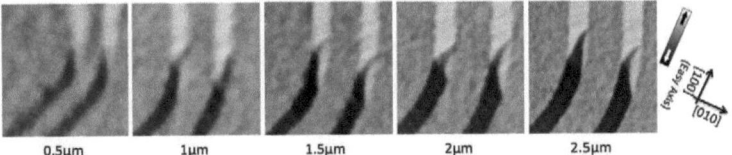

Figure 6.15: Domain walls in LSMO wires with widths from 0.5 μm to 2.5 μm in 500 nm steps. The images are sharpened to adjust the big differences in image quality since the resolution is fixed.

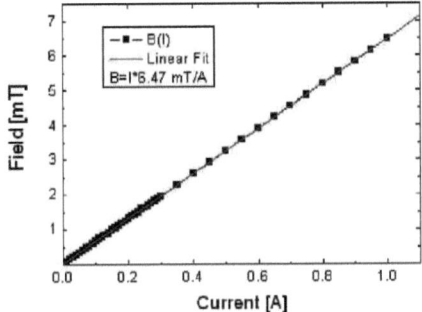

Figure 6.16: Field at the sample position, generated by the sample holder coil.

introduced in the wire bend, which is a prerequisite for transport measurements and it was found that in wire dimensions from 500 nm up to 2.5 μm, transverse walls are present after initialization.

6.4 Field-Induced Nucleation and Depinning

To investigate the response of the patterned LSMO elements to an applied field, the 30 nm EBL pre-patterned LSMO sample was mounted on a PEEM-holder with a mounted coil. The field of the coil was measured by a commercial Hall-sensor, prior to the experiment. The linear behavior of this coil is shown in Fig. 6.16. The remanence of this coil is in the range of 0.1 mT. The field values used in this section are determined by $B = I \times 6.47\,\text{mT/A}$, where I is the applied current. The present domain state that is observed in an element depends on both the history of the sample and the experimental conditions, i.e., the fields and the temperatures that were applied before and during the measurement. In order to separate temperature and field-induced effects, the following experiments are performed at 300 K.

6.4.1 Transformation of Domain Walls and Domain States

To study the influence of an applied magnetic field on the magnetic state of ring elements, a large array of rings with a constant thickness of 30 nm and varying widths was imaged. Images were taken at remanence after applying different fields. These images are summarized in Fig. 6.17 for a 1.5 μm wide ring. The initialized state after magnetizing with -13 mT is presented in Fig. 6.17 (a). The ring is in an onion state and displays transverse domain walls. Next, the field is reversed and an increasing field is applied to the sample. At an applied field of +2.6 mT, one domain wall depins and moves until the two walls annihilate. The resulting vortex domain state is shown in Fig. 6.17 (b). This state remains stable until an applied field of +3.56 mT, where two vortex walls are nucleated and now define a new (reversed) onion state (Fig. 6.17 (c)). Similar experiments were performed in NiFe rings with MOKE (magneto-optical Kerr effect). In the experiments from Kläui et al. [92, 93], hysteresis loops were taken of ring array to obtain the switching fields. This technique allows for a distinction between vortex- and onion state, but a more detailed study of the spin structure is not possible due to the limited resolution. In the experiment performed on the LSMO elements, PEEM was used and in addition to the domain state transition, domain wall transformations due to the applied field are observed (Fig. 6.17 (d)-(g)).

The discussion of the observed phenomena is divided in (I) the transition between the two different domain states, namely the transition from onion to vortex and the transition from vortex to the reversed onion state; (II) the transformation and nucleation of domain walls.

(I) Transition between Domain States

The transition between the onion and the vortex state depends on the depinning of at least one of the domain walls, which is a stochastic process that depends on local pinning sites and thermal activation. The number of the vortex state transitions for different ring widths is given in Fig. 6.18. The fraction of switched states is given in % and is plotted as a function of the applied field. One can see a clear dependence of the transition field on the ring width. The first onion state of the widest element, with a width of 1.8 μm (black), already switches into the vortex state at a field slightly above 1 mT, which is close to the coercive field of this film (determined on a large epitaxial area). The field values for the transi-

6.4. Field-Induced Nucleation and Depinning   107

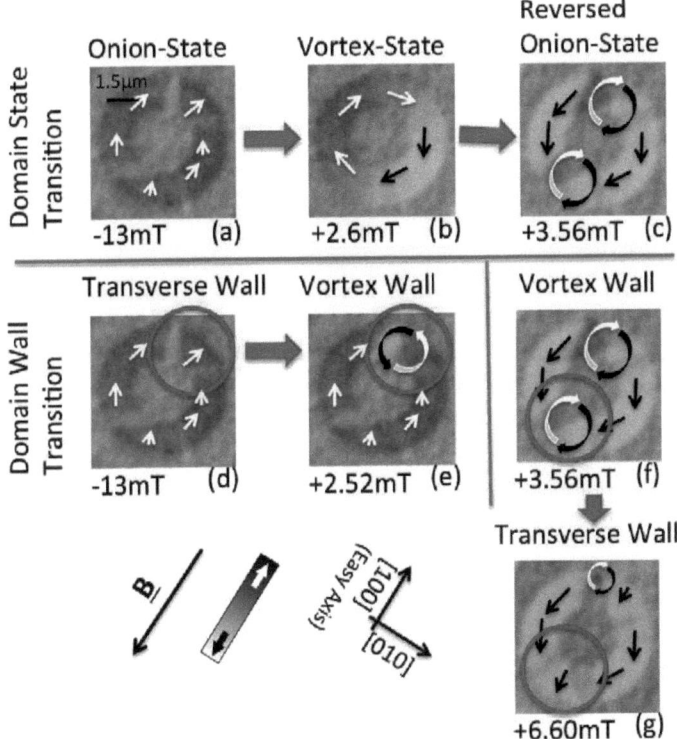

Figure 6.17: Transition of domains and domain walls in a 1.5 μm wide and 30 nm thick ring shaped element by an applied magnetic field. The saturated state in (a) shows an onion state, which switches into the vortex state in (b) at an applied field of 2.6 mT. After further increasing the field, domain walls are nucleated and the vortex state transforms into a reversed onion state. One of the initial transverse walls (marked with the red circle) transforms into a vortex wall at 2.52 mT ((d) and (e)). (f) shows the onion state after the domain state transformation, which shows vortex walls. At a field of 6.6 mT, one can see that the wall transforms into a transverse wall (g).

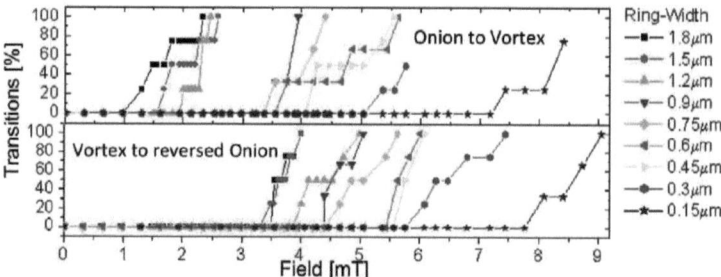

Figure 6.18: Transition between domain states in 30 nm LSMO. The top graph shows the transition from the onion to the vortex state in % of the observed domain states. The transition from the vortex state into the reversed onion state is shown in the bottom graph.

tion are spread in a range over 1 mT, which is a sign that the transition is not only determined by the geometry of the element but also influenced by random pinning. The switching field increases continuously with decreasing width, which goes down to 150 nm (dark blue) in this experiment. At this width the onion states remain stable up to applied fields of above 7 mT. At this field all the other elements have already switched.

As shown in Fig. 6.17, the vortex state switches to a reversed onion state after further increasing the field. The lower graph in Fig. 6.18 shows this transition and one can again see the width dependence of the transition.

From these data, the mean switching field for every ring geometry is determined by extracting the field that is needed in order to switch 50% of the states by linear fitting the curves in Fig. 6.18 in the region of the transition field. The resulting distribution is presented in Fig. 6.19. The black curve is the transition of an onion state into the vortex state, the red curve is the transition to the reversed onion state. Similar experiments on the transition of domain states in Co rings are available in the literature [94, 95].

The increasing switching field from onion to vortex state for decreasing ring width was predicted in [96] and experimentally demonstrated in [93]. The same characteristics is observed in this experiment and, interestingly, the difference between the fields for the onion-vortex and the vortex-reversed-onion decreases nearly to zero (marked with a red circle in Fig. 6.19 (a)) at narrow elements. A possible explanation could be the nucleation field for a domain wall, which is en-

6.4. Field-Induced Nucleation and Depinning

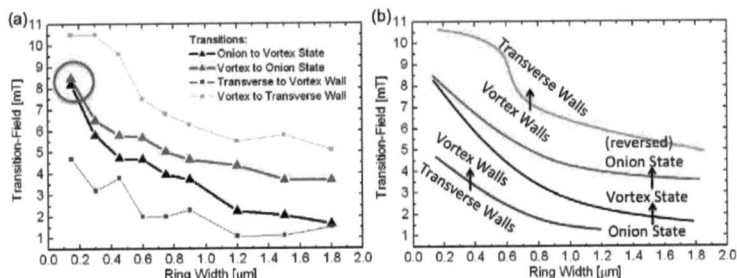

Figure 6.19: Switching of vortex and onion state in relation of applied field. (a) Experimental results and (b) idealized sketch to illustrate the findings.

ergetically above the vortex to onion state transition for wide elements but leads to the same (or even less) value for narrow elements.

(II) Domain Wall Transformation and Nucleation

Before the domain state switches from the onion into the vortex state, some of the transverse domain walls transform into vortex walls. The depinning field of vortex walls is lower than the depinning field of transverse walls [53]. This can lead to a displacement of the vortex domain wall, followed by the annihilation of the two domain walls. This displacement can even occur before both domain walls transform. Examples of domain wall transitions are given in the bottom row of Fig. 6.17. In Fig. 6.17 (d) and (e), the transition of a transverse wall into a vortex wall at 2.52 mT is shown (labeled with the red circle).

The transition of a vortex wall into a transverse wall in the reversed onion state is shown in Fig. 6.17 (f) and (g). Here, the vortex core is displaced until it is at the border of the element. The field and size dependence of these transitions is given in Fig. 6.19. The blue line represents the mean transition field from transverse walls into vortex walls. The green line is the transition from vortex walls into transverse walls. The graph on the left side contains experimental data, the graph of the right side is a sketch, which contains the main characteristics. We observe that the first transition from transverse into vortex wall (blue line) appears approximately at half the field for the depinning field of the domain wall (black line). A transverse wall has a high stray field and is energetically not favored over the vortex wall for wide rings, but the strong pinning at the edge is able to trap this domain state. Both transverse walls and vortex walls are stable spin configurations in all these ring elements at room temperature. The formation

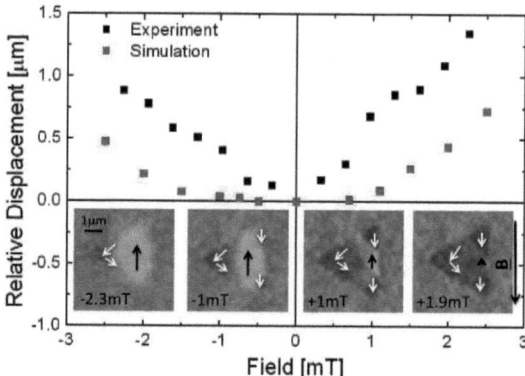

Figure 6.20: Vortex core of a 30 nm thick EBL defined LSMO triangle, displaced by a magnetic field in experiment and simulation. Insets are XMCD-PEEM images at various fields. The displacements of the vortex core can clearly be seen. The field is applied in the horizontal direction.

of a vortex core needs energy, which can be provided by the spin structure itself, the applied magnetic field, and also thermal activation.

After the vortex state transforms into the reversed onion state, mostly vortex domain walls are present. Further increasing of the field leads to a vortex core displacement inside the domain wall. The total vortex core displacement for a wide ring is larger than for a narrow ring, but the potential wall is less steep. The green transition curve in Fig. 6.19 shows that the vortex walls in wide elements switch into transverse walls at lower fields than those in the narrow elements.

For very narrow elements, we see large fields for the transition from vortex to transverse wall (Fig. 6.17, green line) and for elements below 0.4 μm, the size dependence becomes less pronounced.

6.4.2 Field-Induced Vortex Core Displacement

The position of a vortex core in a constrained element is determined by the local potential. For small displacements, the potential landscape can be detected for example by homodyne detection [50]. For larger displacements and complex shape geometries, spatially resolved imaging of a domain state in a field can give information about the potential landscape. An example for a non rotational symmetric potential for a vortex core is the potential landscape of a magnetic triangle.

6.4. Field-Induced Nucleation and Depinning

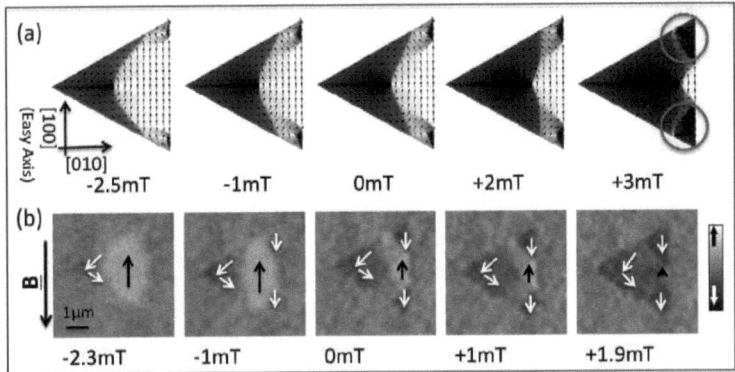

Figure 6.21: Comparison of numerical simulations with the experiment. The triangles of the numerical simulations in (a) show a similar spin structure like the experimentally observed 30 nm thick EBL defined LMSO triangles in (b). Different field values are applied and deform the spin structure until the centered vortex is pinned at the right edge at large field values.

In Fig. 6.20, different vortex core positions in a triangle in varying magnetic fields are shown. In order to generate these Landau states in triangular elements, the sample was saturated with a strong magnetic field and after applying a smaller field in opposite direction, the states switched to simpler magnetic states. In the 30 nm thick LSMO film, the domain states were present in complex Landau states. This makes the precise analysis of the potential landscape for the vortex core difficult, as the different domains grow and shrink in size, depending on the direction of the applied field. The imaging of the domain states with displaced vortex cores was performed with an *in situ* applied field.

The electron based imaging technique of the PEEM shows a deflection of the image when a magnetic field is applied. Up to field values in a the range of 1-3 mT, this translation can be compensated with the object alignment system of the PEEM. The distortion of the image and the loss of sharpness are compensated with the objective and the stigmator lenses. A drastic loss of image quality is observed when imaged at large applied fields, as seen in the Fig. 6.20 (left).

In this example, the triangular element consists of a large Landau state with a well pronounced vortex core in the center and two smaller (dark) areas in the two right corners. Nevertheless, the observed potential must be asymmetric along the horizontal direction. The central vortex core moves towards the left corner for negative fields with growing domains in the two right corners. A positive field pushes the centered vortex core to the right edge with shrinking domains in the right corners. The displacement of the central vortex core with respect to the applied field is shown in Fig. 6.20. The displacement in positive field direction, where the core is pushed towards the edge, is larger than the displacements in a negative field. The main origin of this behavior can be attributed, again, to the stray field energy, which becomes larger when the spin structure is forced into the triangle vertex edge at the left side of the element. The displacements are reproducible, which was confirmed by successively measuring remanent position of the vortex core, which only varies in a small negligible range.

In order to be able to repeat the field displacement of the vortex core several times, the field values in negative field direction are only applied up to values that do not lead to an annihilation of the vortex core. For positive field directions, the field was increased until the vortex was annihilated on the right side, resulting in an irreversible "frozen" domain state (see Fig. 6.20, right). Micromagnetic simulations are, again, performed and compared with the experiment (see Fig. 6.21).

While one can see vortices in the right corners of the simulated configurations, the experiment shows instead a uniform dark contrast in this region, indicating a uniform spin orientation. The origin for this could be the strong pinning in this LSMO film, induced by the polycrystalline area around the element. This prevents the vortex core from nucleating, which is not the case in the idealized simulation. A pinning of the small vortices in the simulation was achieved at higher field, as seen in Fig. 6.21 (a) at a field of 3 mT (marked with red circles). In the simulation, we find a relationship between the vortex core position and the applied field that is similar to the experimentally observed vortex core position (Fig. 6.20). However, the vortex core displacement in the simulation is slightly smaller than in the experiment. This can be attributed to the parameters used in the simulation, which may differ from the real parameters and lead therefore to the slightly different spin structure.

6.4.3 Field-Induced Domain Wall Depinning and Displacement

Domain wall displacement is influenced by pinning at edge defects and by local pinning sites. For devices, based on the translation of domain walls, reproducible domain wall motion and low depinning fields are required. In order to determine the depinning field of wires in the 30 nm EBL patterned LSMO film, XMCD-

6.4. Field-Induced Nucleation and Depinning

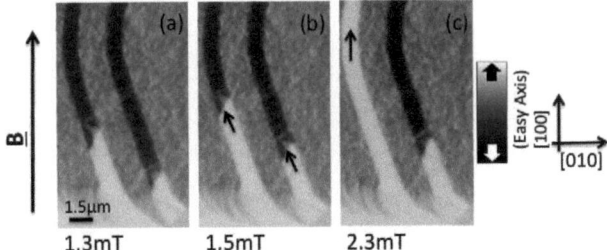

Figure 6.22: Displacement of domain walls in 1.5 μm wide and 30 nm thick EBL patterned LSMO wires.

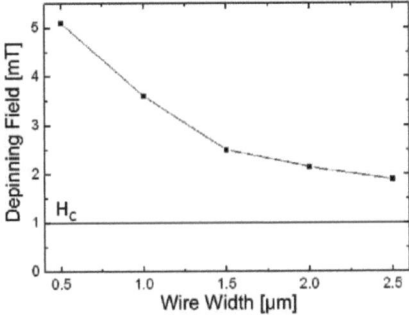

Figure 6.23: Depinning field for different wire widths in 30 nm LSMO. The field decreases with size and will converge to the coercive field H_c of the continuous LSMO film for very wide wires.

PEEM images of the wires are shown in Fig. 6.15 of Section 6.3.2. These wires show transverse walls after magnetizing against the bend along the [010] direction.

In this experiment, fields are applied along zig-zag wires in [100] direction. The widths vary between 0.5 and 2.5 μm. The patterned wires are directly connected to a larger LSMO pad. A sequence of images at remanence after a field was applied is shown in Fig. 6.22. Here, two 1.5 μm wide wires are shown during the magnetization process. Prior to this experiment, the sample was magnetized in negative field direction. The big pad was magnetized in positive field direction (white contrast) at around 1 mT which equals to the coercive field of the epitaxially grown LSMO film. The field needed in order to push a domain inside the wire is above this value.

Some wires have a broken connection to the large pad. In this case, the domain wall needs to be nucleated. This nucleation required larger fields than the field needed for the depinning of a wall and therefore these fields are not taken into account for the analysis of the depinning field.

At an applied field of 1.3 mT, both domains have entered the wire. Instead of transverse walls, which have been observed in the same wires in Section 6.3.2, mainly vortex walls are observed in this experiment when a wall is pushed into the wire. The previously observed transverse walls in Section 6.3.2 can be explained by edge induced pinning (Section 6.4.1).

By further increasing the field, the domain walls depin and move along the field direction. We observe that local pinning sites dominate the propagation. The example in Fig. 6.22 shows a wall that already moved out of the field of view at a field of 2.3 mT (left wall) and one that is still trapped at a pinning site close to the pad. The depinning field as a function of different wire geometries is given in Fig. 6.23. Here an average value was chosen, where approximately 50% of the wires have switched. Smaller wire dimensions require larger magnetic fields to depin the domain walls analogous to the observed depinning in rings (see Section 6.4.1). This behavior for domain walls was previously observed by Adeyeye et al. in Permalloy [97, 98], where a s/w dependence for the depinning field was found. Here, w is the element width and s the spacing between different wires of an array. This $1/w$ dependence agrees well with the depinning fields in LSMO domain walls that are shown in Fig. 6.23. Here the spacing between the wires is relatively large and a direct coupling is not expected.

6.5 Thermally Activated Effects in LSMO

In micromagnetic theoretical descriptions of the spin configuration of magnetic elements, one often simulates at T=0 K. Simulations at temperatures above T_0 are more complex and very time consuming. However, real experiments are always conducted above absolute zero and only cryostat experiments, cooled with liquid helium come close to this temperature. However, by adjusting the micromagnetic parameters accordingly, one can simulate the various observed effects and the domain states numerically.

When a magnetic material is heated up, the saturation magnetization will be reduced. When the temperature reaches T_c, the magnetic order vanishes completely and the material undergoes a transition into the paramagnetic state. This

6.5. Thermally Activated Effects in LSMO

is interesting insofar as it allows one to change the magnetic properties gradually *in situ*. Application of heat-induced effects are already used in magnetic storage devices (for example the "Sony MD"). Here, the magnetic softening of a heated ferromagnet is used to magnetize a well defined area.

Studies of the temperature influence on domains in Permalloy can be found in [90]. Here, the magnetic system was heated up and imaged up to temperatures above 600 K. These temperatures can be easily generated in commercially available PEEM heating holders. The heat is generated by a tungsten filament and the sample temperature measured by a thermocouple. If higher temperatures are needed, electron beam heating can be used. The energy of this electron bombardment can be tuned to heat the sample to temperatures above 1300 K but such high temperatures may lead to micro-structural changes.

Fortunately, the Curie temperature of LSMO is not too high above room temperature, which makes it easy to image the changes of the spin structure during the heating process, without the drawback of drift problems. Also the heating and cooling times are strongly reduced if only moderate temperatures are used. In the following, different experiments on FIB patterned LSMO and EBL pre-patterned LSMO elements with different thicknesses are presented.

In Section 6.5.1, results of domain states are presented where the temperature is slowly increased to above T_c to study the thermally activated change and nucleation of domains. In Section 6.5.2, domain walls are heated up and the transition between the different domain wall types, analogue to the experiment of Laufenberg et al. [90], is observed.

6.5.1 Thermal Depinning and Freezing of Domains

In order to investigate temperature induced effects in LSMO, the samples were mounted on heating a PEEM holder with a tungsten filament and a thermo couple.

The heating sequence of a large square element with an edge size of 12 μm is shown in Fig. 6.24. In (a), the element is in a relaxed state with relatively large domains. The grains of the surrounding polycrystalline material affect the edge roughness, which is visible by the small distortion of the magnetization starting from the edge of the element. The XMCD contrast in 60 nm thick sample is very high and also present in the surrounding grains, which means that they are ferromagnetic albeit with very small or even mono domains.

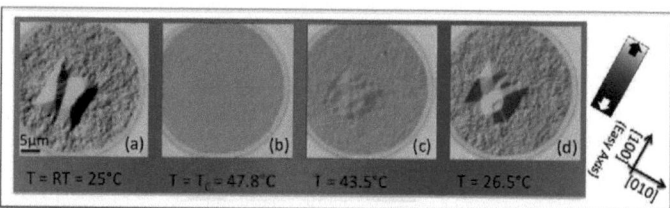

Figure 6.24: Series of XMCD images of a 60 nm thick EBL patterned LSMO film. The temperature is increased from room temperature (a) to T_c (b). Upon cooling down from temperatures above T_c, small domains nucleate (c). During the cooling process, the domains combine and form larger domains (d). Note that the domain states in (a) and (d) are different at similar temperature.

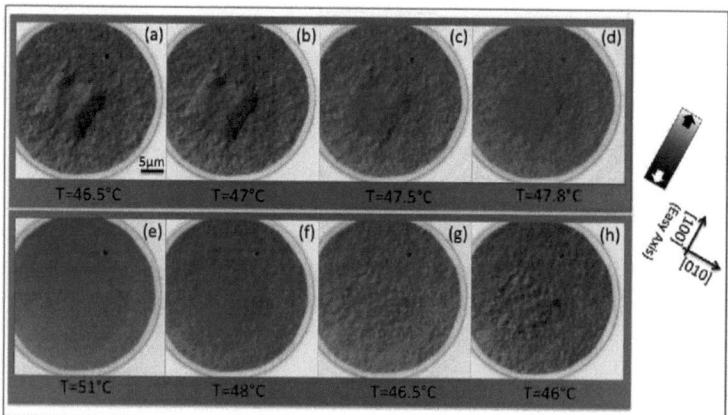

Figure 6.25: Contrast enhanced XMCD images, taken on a 60 nm thick EBL patterned LSMO film around the critical temperature T_c. The initial domain state is shown in Fig. 6.24. In (a), this initial domain state is still identifiable at a temperature of 46.5°C. (b) shows the slowly degenerating domain state at 47°C. At 47.5°C (c), the central part of the square element looses its magnetic contrast. The magnetic contrast of the patterned element vanishes completely at 47.8°C (d). The polycrystalline area around the square remains magnetic up to a transition temperature of 51°C (e). After cooling down to 48°C (f), the magnetic contrast returns in the polycrystalline area. (g) and (h) shows the cooling process with small nucleated domains, that combine to larger domains during cooling.

6.5. Thermally Activated Effects in LSMO

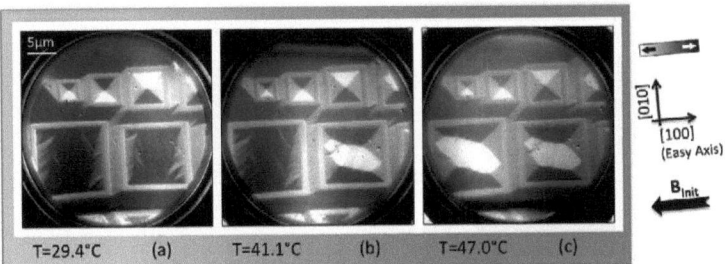

Figure 6.26: XMCD images of 50 nm FIB patterned LSMO films. (a) the small elements are in a flux closure spin structure. The two biggest elements are in a frozen C-state. (b) At a temperature of 41.1°C, the second smallest structure falls into a complex flux closure domain state. At 47°C, also the biggest structure transforms into a complex flux closure spin structure (c).

At a temperature of 47.8°C, the contrast in the square element vanishes completely, whereas the contrast in the surrounding grains becomes weaker, but remains (b). The measurements indicate a loss of magnetic contrast close to T_c at a temperature close to 48°C. After cooling down below T_c, small domains form (c) and reorientate during the cooling process (d). The contrast enhancement in this series of images is the same for all images.

To have a more detailed look inside the processes close to T_c, the contrast in that temperature regime is further enhanced in the image series shown in Fig. 6.25. In this image sequence, one can see the collapse of the domain state close to T_c. At a temperature of 46.5°C (Fig. 6.25 (a)), the domain state is still comparable to the original state in Fig. 6.24 (a). Smaller domains have already formed in the element due to local fluctuations and by increasing the temperature another 0.5° to 47°C, the domain boundaries vanish (Fig. 6.25 (b)). At 47.5°C (Fig. 6.25 (c)), the contrast in the central part of the patterned element has lost contrast and appears uniformly gray.

With this technique and with exposure times in the order of one minute, one cannot resolve possibly existing domains that are not localized anymore. Only the area close to the borders still shows some contrast and also the polycrystalline area around the element remains ferromagnetic. This could be attributed to the exchange that keeps the small polycrystalline grains in a monodomain state and makes thermally activated switching less likely. Therefore the fluctuations are shifted to a larger timescale and imaging of the magnetic contrast is possible very close to T_c. This exchange "fixes" the domain state at the border of the element,

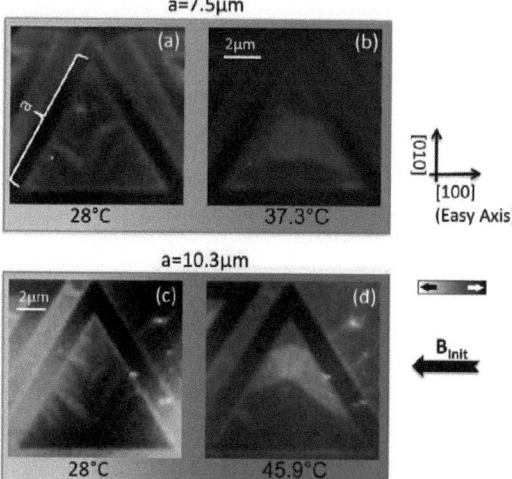

Figure 6.27: A 7.5 μm large and 50 nm thick FIB patterned LSMO triangle shows a pinned spin structure after magnetization (a). The spin structure forms an energetically favored flux closure state at transition temperature of 37.3°C (b). A triangle with an edge length of 10.3 μm shows a similar transition to a flux closure state at 45.9°C (c)-(d).

6.5. Thermally Activated Effects in LSMO

thus allowing for imaging the spin configuration that is at least stable over some minutes.

At 47.8°C (Fig. 6.25 (d)), the Curie temperature for the epitaxially grown area is reached. The T_c for the polycrystalline area is slightly higher and at 51°C (Fig. 6.25 (e)), the contrast vanishes completely. When the sample is cooled down again, the contrast first returns at the polycrystalline area (Fig. 6.25 (f)). As cooling is continued, very small domains form in the element (Fig. 6.25 (g)). Here, the randomly distributed spins of the paramagnetic phase start to freeze and all over this area small domains are formed independently.

As the temperature decreases another 0.5°C, the small domains transform into bigger domains, in order to minimize the domain wall energy (Fig. 6.25 (h)). At these temperatures close to T_c, the whole domain state is dominated by random thermally activated effects and not by the shape of the element. The changes happen quite rapidly from image to image. The consequence of this is a loss of the average contrast. Fast (of the order of the exposure time ($\sim$1 min)) moving and transforming domain walls may also contribute to the vanishing contrast in the element. This suggests that the real T_c may be higher than the temperature at which the contrast is lost. The small monodomain states in the polycrystalline area need much higher energies to fluctuate and therefore show a contrast up to temperatures very close to T_c.

In Section 6.3.1, the results at room temperature show that flux closure domain states are present in FIB patterned LSMO elements up to a few micrometers in lateral size. Larger elements show "frozen" domain states. These energetically unfavored spin configurations are stabilized by pinning of the spin structure at the element edge. This pinning can be overcome by thermal activation.

In Fig. 6.26, elements with different sizes in 50 nm FIB patterned LSMO are shown. The domain states show flux closure domains up to element dimensions of 4-5 μm. The two biggest elements have an edge length of 10 μm and 12 μm, respectively, and have a C-state spin structure. The 10 μm large element falls into a flux closure spin structure at 41.1°C. The 12 μm size element is less influenced by the shape and transforms into a flux closure spin structure at a temperature of 47°C.

The same behavior can be observed in triangular elements. The triangles shown in Fig. 6.27, show a transition from the "frozen" into the flux closure state at 37.3°C for the 7.5 μm large element (a)-(b) and a transition at 45.9°C for the 10.3 μm large element (c)-(d). The higher transition temperatures for large

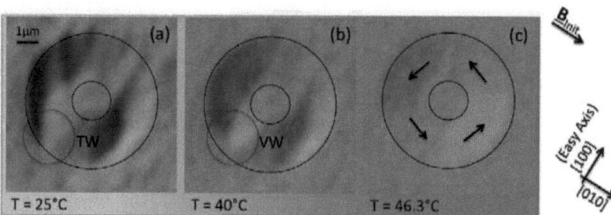

Figure 6.28: XMCD images of a 1.5 μm wide ring element in 30 nm thick EBL patterned LSMO. The initialized onion domain state in (a) shows two transverse walls at room temperature. A vortex core nucleates at a transition temperature of around 33°C in the left domain wall at the outer border of the ring. During further heating, the vortex core moves towards the center of the domain wall and the state at 40°C is shown in (b). Further heating leads to further changes in the spin structure and an annihilation of the two domain walls is observed, resulting in a flux closure vortex state (c).

elements are caused by the weaker influence of the shape.

6.5.2 Thermally Activated Transformation of Domain Walls

In field-induced experiments, it was shown that both transverse and vortex walls can be stable in the 30 nm thick EBL patterned LSMO film (see Section 6.4.1). Heating of a domain wall can change its spin structure and as already demonstrated in [90], a transition of a transverse wall into a vortex wall can occur. In our experiments, transverse walls can be generated in rings by saturating the sample. The induced onion states mainly show transverse-like domain walls. That vortex walls are also stable in this material for narrower element dimensions was already demonstrated in Section 6.4.1 and 6.4.3.

The heating process of a 1.5 μm wide element is presented in Fig. 6.28. The transverse wall in (a), with the typical triangular shape, is marked by the small red circle. At the transition temperature around 32.5°C, the wall transforms into a vortex wall (b). At higher temperatures, more transformations inside the rings take place and some domains displace and finally annihilate. This annihilation leads to a formation of a vortex state in the ring (c), shortly before reaching T_c.

The transformation of the domain walls from transverse into vortex as a function of ring width is shown in Fig. 6.29. Surprisingly, we cannot see a dependence of the nucleation temperature on the ring width for elements in the range of 1.8 down to 1.2 μm. This lack of dependence on the ring width can originate from

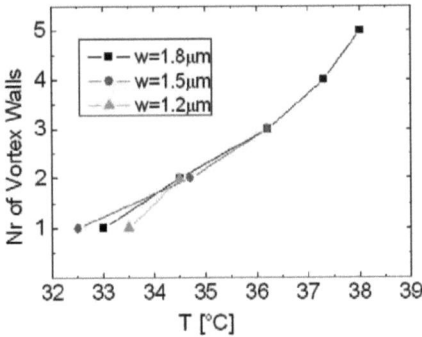

Figure 6.29: Number of thermally activated transition of transverse domain walls into vortex walls for 30 nm thick rings in EBL patterned LSMO for ring widths from 1.2-1.8 µm. Five rings with two domain walls each are observed. The domain wall transitions start around 33°C and do not show a significant dependence on the element width.

the pinning of the transverse wall at the element edge. The strength of the pinning sites are independent of the ring geometry and can be overcome at similar temperatures for different elements.

Transverse walls are stabilized in narrow rings, so a transition at higher temperatures in narrow rings would be expected. The randomly distributed pinning sites and the small statistics do not allow this to be determined in the current experiment. Smoother wire edges and a wider range of investigated sizes would be required to investigate the temperature induced domain wall transformation in more detail.

6.6 Conclusion

In this chapter, results were presented on domain structures in patterned half-metallic LSMO thin films. Two different methods, one with FIB and one with EBL pre-patterned Cr seed films were developed to pattern different micro- and nano sized elements. Studies on triangular and square elements, together with SQUID measurements indicate a small uniaxial anisotropy in the films, which is much smaller than reported in the literature [99, 100].

In particular, the role played by this anisotropy term are seen in the comparison of domain states of unrotated states and of 45° and 90° rotated elements. A uniaxial anisotropy in LSMO can be induced by a lattice mismatch to the substrate or by periodic crystalline steps [45]. However, compared to similar experiments [99, 100], flux closure domain states that are mainly determined by the shape anisotropy with only a small influence from the uniaxial anisotropy were observed, which shows the high quality of the films with uniform strain or small epitaxial step induced magnetocrystalline anisotropy.

Domain walls were found in the in-plane magnetized half-metallic structures. The domain wall spin structures in wide wires are more complex than in narrow wires. Imaging of domain states at different angles allow for a full analysis of the spin orientation. By applying magnetic fields, changes of the spin structure are imaged. By magnetizing small ring elements, the transition of domain walls and also the transition of domain states are determined as a function of the magnetic field.

The imaging of field-induced domain wall motion in LSMO nanowires show the dependence of domain wall depinning with respect to the wire width. The large variation of the depinning field for a given geometry depicts the strong influence of local pinning sites.

The thermally activated studies show the different behavior of polycrystalline and epitaxially grown LSMO. In particular, the pinning sites that keep a domain state in an energetically unfavored state can be overcome by thermal activation. The effect of thermally activated depinning is also the origin for domain wall transformation from transverse into vortex walls in ring elements. Heating experiments on epitaxially grown LSMO elements show the change of the spin configuration close to T_c.

In conclusion, the spin structures in half-metallic LSMO, which might have a future in spintronic applications, can be tuned by the appropriate choice of design and film thickness. Films showing a very small magneto-crystalline anisotropy are good candidates for spin injectors, as the spin configuration can be easily modified by geometry, field, and temperature. With Curie temperatures above room temperature, real device applications are feasible, although the fact that the Curie temperature is around 340 K, might be a little low for real applications involving spin-torque effects. Nevertheless, a low T_c makes it more accessible for thermally induced experiments, and allows for an easier analysis of temperature-induced effects.

Future experiments on this system could involve the fabrication of structurally isolated elements. These could be patterned by e-beam lithography on extended

6.6. Conclusion

films, using negative photoresist and ion milling (see Section 3.2.2). A possible reduction of the edge roughness could result in low critical current density ies for current-induced domain wall motion [9]. In spin-valves, the LSMO could improve the spin current and the spin valve signal due to its high spin polarization [88].

CHAPTER 7

Spin Configuration in Patterned Heusler Alloys

Heusler alloys are a special type of material structure and some belong to the class of ferromagnetic half-metals with one semi-conducting and one metallic spin band (see also LSMO, Chapter 6). X_2YZ Heusler alloys crystallites in highly (110)-oriented $L2_1$ structure [101], consisting of four fcc sublattices and can exhibit half-metallicity [102–104]. A half-Heusler compound is predicted to have metallic spin band and one spin band with a gap [105]. This leads to rich electronic behavior with a very high spin polarization at the Fermi-level (theoretically up to 100% [79]) and makes this class of materials a possible candidate for future spintronic devices and applications [7, 10]. Co based Heusler alloys have a very high Curie temperatures (e.g. T_c=985 K for Co_2MnSi) and very large magnetic moments up to 5.5 μ_B per unit cell (e.g. Co_2FeGe) [106].

To utilize these interesting features, a detailed knowledge about the spin structure in these materials is necessary. Photoemission electron microscopy (PEEM, see Section 2.1.1) and magnetic force microscopy (MFM) experiments are performed in order to reveal the spin structure of EBEAM patterned nano- and micrometer sized Heusler alloy elements.

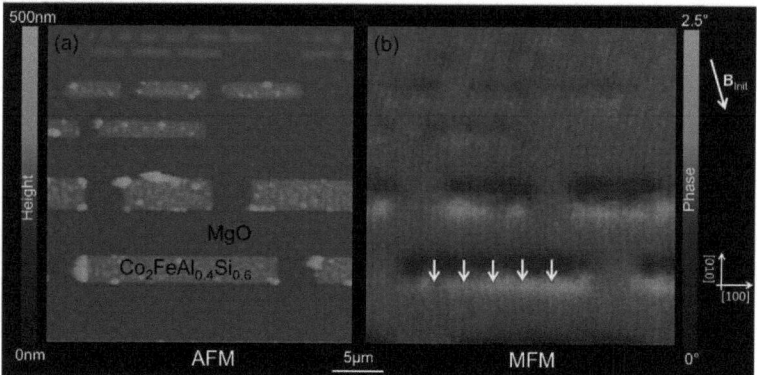

Figure 7.1: (a) AFM image of 30 nm thick patterned $Co_2FeAl_{0.4}Si_{0.6}$ alloy. The corresponding MFM image with magnetic contrast is shown in (b). A large difference in phase indicates a stray field (bright-dark contrast), which is generated by large magnetized areas that are not present in a flux closure state.

7.1 Experiment

In this work, the Heusler alloy $Co_2FeAl_{0.4}Si_{0.6}$, which in order to grow half-metallic $Co_2FeAl_{0.4}Si_{0.6}$ requires an appropriate seed layer. The lattice constant of bcc Cr provides the necessary base to form the $L2_1$ phase. For $Co_2FeAl_{0.4}Si_{0.6}$, post annealing is required to induce strong (001) texture and form the Heusler phase.

The Cr seed layer was grown on single crystal MgO substrates. After deposition, the alloy is annealed for 30 min at 500°C in either vacuum or nitrogen atmosphere to form the (100) texture. Since the deposition and annealing can be done independently, EBL patterning can be used to pattern non-ordered $Co_2FeAl_{0.4}Si_{0.6}$ and to form the Heusler phase afterwards. This fabrication process makes it easy to structure well defined micro- and nanometer sized elements.

For magnetoresistance measurements and for initial characterization, Heusler alloy structures are patterned on MgO. The magnetoresistance measurements reveal interesting magnetic transport properties (for instance the existence of the planar spin-Hall-effect [107]). SEM images of the samples used for the magnetoresistance measurement are given in Fig. 3.9.

To understand and interpret the resistivity data, a knowledge of the spin structure in these confined elements is crucial. AFM/MFM images on the same sample but on different elements are shown in Fig. 7.1 and Fig. 7.2. The topological

7.1. Experiment

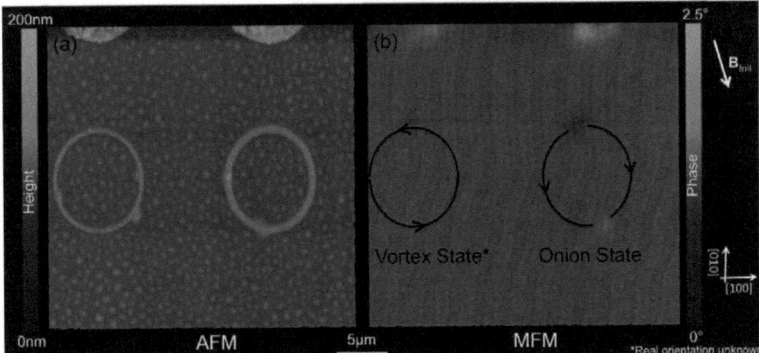

Figure 7.2: (a) AFM image of two 30 nm thick $Co_2FeAl_{0.4}Si_{0.6}$ rings. The stray field, visualized by an MFM measurement is shown in (b). The right ring shows two areas with rich contrast. One is brighter and one is darker than the background. This contrast indicates the presence of domain walls. The left ring does not show a stray field, which means that the spin configuration of the ring is in a flux closure vortex state.

contrast in these images is shown in (a) and the corresponding magnetic MFM contrast is shown in (b). The rectangular elements in Fig. 7.1 (b) show a large contrast at top and bottom of the element, indicating a magnetization along the [010]-direction, which can be attributed to a large magnetic anisotropy.

The magnetic contrast of Fig. 7.2 (b) shows that, after magnetizing, domain walls are present in rings. Here, the left ring is in a vortex state and the right ring is in an onion state, thus showing two domain walls.

The insulating MgO, which is appropriate for transport measurements, is problematic for PEEM measurements and EBL due to charging issues. This problem was solved by patterning the structures on a continuous Cr thin film, which is deposited on the whole sample surface prior the EBL.

A second approach to pattern a thin film is the spatial removal or structural change of the film by focused ion beam lithography (Chapter 3.1.3). Fig. 3.3 shows an XMCD image of a FIB patterned area, together with the topological absorption image. Although the FIB defined elements show a different contrast, the stripe domains are clearly visible.

7.2 Spin Configuration in Basic Shapes

The MFM images in Fig. 7.1 and Fig. 7.2 already prove the feasibility of creating magnetic micro- and nanometer-sized Heusler alloy elements, but they do not allow for a more detailed insight into the spin structure, as this technique is only stray field sensitive and therefore not ideally suited for in-plane magnetized materials. Therefore, in order to get a basic understanding of domain sizes and domain walls in Heusler alloys, basic shapes are investigated by PEEM.

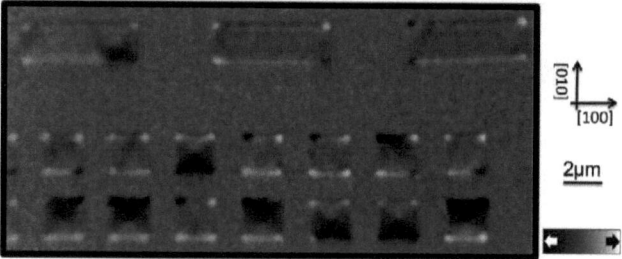

Figure 7.3: XMCD-Photoelectron emission microscopy (XMCD-PEEM) image of squares and rectangles. Half of the squares show flux closure states, the other half is in a less defined state. The rectangles mainly show ripple domain structures.

7.2.1 Spin Structure in Squares, Rectangles and Disks

In Fig. 7.3, the XMCD contrast of EBL patterned squares and rectangles are shown. One can see that 50% of the squares show a flux closure Landau pattern, whereas the other half are in less well defined states, with a weak magnetic contrast. Only at the corners, a strong contrast is visible, probably originated by the different annealing at the edge regions. Here the Cr interdiffusion or the crystal lattice have an influence that seems to locally increase the magnetic order. This behavior points to a very strong uniaxial anisotropy that overcomes the shape anisotropy and leads to the formation of non flux closure spin structures. In the bar elements at the top of Fig. 7.3, the same effect in the corners can be observed.

Figure 7.4 shows disks with varying size from $1\,\mu$m to $3\,\mu$m. As the disk size increases, the domain states go from mono- and bi-domain states in the smallest disks to more complex states in the largest disks. Another, not fully understood, phenomena is the increase of the XMCD contrast for smaller element sizes. The

7.2. Spin Configuration in Basic Shapes

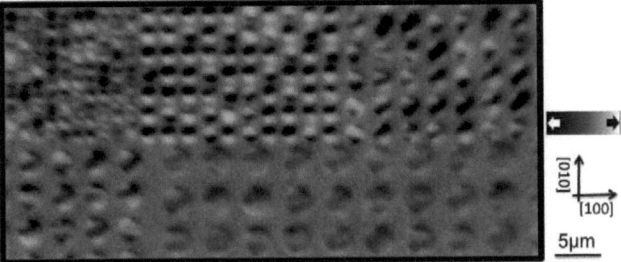

Figure 7.4: XMCD-PEEM image of disks with varying size. The smallest disks show single and double domain states. The bigger disks show double and triple domain states. The XMCD contrast increases with decreasing size.

size dependent XMCD contrast was found throughout all experiments. The larger contrast in small elements could be explained by Cr interdiffusion from annealing or by an out-of-plane tilting of the spins in the larger elements.

The competing anisotropies only allow a small range of element sizes to have a shape induced spin configuration. It was found that elements with typical dimensions in the range of 500 nm show both a strong magnetic contrast and a well defined and reproducible domain structure.

7.2.2 Domain Walls in Rings and Nanowires

For domain wall based applications and experiments, the existence of domain walls is a prerequisite. To study domain walls in Heusler alloy elements, rings and zig-zag shaped wires were imaged.

Fig. 7.5 and Fig. 7.6 show rings after magnetizing with a strong field along the [010]-direction. The resulting onion states show transverse domain walls that divide the two quasi uniform domains. A strong tendency for ripple domains in the large elements and an enhancement of the XMCD contrast in smaller elements and at edges is visible. The spin configuration of these onion-states are similar to the ones observed by Fonin et al. [108], pointing to a four-fold magnetocrystalline anisotropy. A better defined domain structure for the smaller element sizes is observed in Fig. 7.6 and can be explained by the influence of the shape anisotropy, that increases for smaller element dimensions (similar to the results of Section 7.2.1). The best reproducibility is achieved for an element width of 500 nm. For this element width, both transverse and vortex walls are observed,

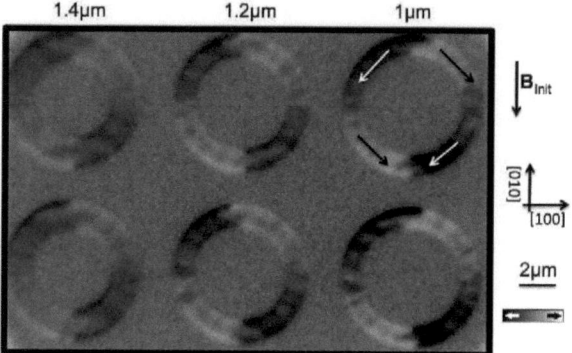

Figure 7.5: XMCD-PEEM image of three different ring widths from 1-1.4 μm is shown. The sample was magnetized prior to imaging. All rings are in an onion state. The spin structure is dominated by ripple domains and the two induced 180° domain walls. The XMCD contrast increases with decreasing ring width.

pointing out that both spin configurations are stable states at room temperature.

Fig. 7.7 shows a series of zig-zag wires that are magnetized along the [010] direction. The shape anisotropy leads to head-to-head and tail-to-tail spin configurations at the wire bend, so that domain walls are formed. The element width is 1 μm and therefore in the range of shape defined spin structures, but the vanishing contrast and the ripple structure is already visible. For transport measurements, wires with a slightly narrower width would be more suited.

The results show that domain walls in Heusler alloys can be generated reproducibly. The XMCD contrast increases for smaller structures and the domain wall spin configuration becomes more complicated for wider elements. In elements with typical dimensions around 500 nm, the spin configuration of the domain walls in the patterned rings are mostly determined by shape anisotropy.

7.3 Thermal Effects

The influence of temperature on the magnetic configuration was previously discussed for patterned LSMO thin films (see Section 6.5). The Curie temperature of LSMO is close to room temperature and therefore easy to access without the drawback of thermal drift problems. The temperature needed in order to influence

7.3. Thermal Effects

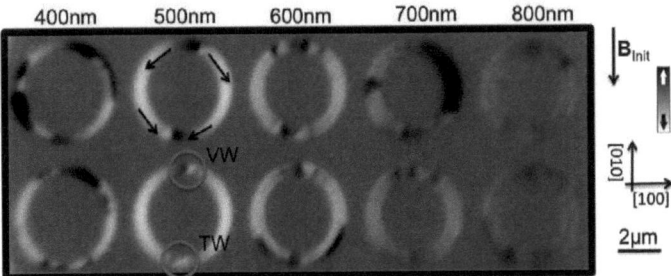

Figure 7.6: XMCD-PEEM image of rings with varying width. There is only a small range of ring widths where a well defined onion state is observed. In this case, the second and third narrowest show the most reliable domain configuration. The narrowest rings fall into multidomain states, while the widest elements show onion states with ripple domains.

the investigated Heusler alloy is larger, but still accessible with PEEM heating holders.

7.3.1 Experiment

To investigate the temperature activated effects in $Co_2FeAl_{0.4}Si_{0.6}$, the sample was mounted on a PEEM heating holder that has a built-in filament and a thermocouple to determine the temperature. To reach high temperatures, which were needed for this experiment, e-beam heating was used.

The heating experiment concentrates on domain walls in ring geometries with varying sizes, from $0.28\,\mu m$ to $0.56\,\mu m$. After magnetizing, transverse domain walls are present in the rings. Heating of the sample and subsequent imaging allows on to study domain wall transformations. Examples of transverse to vortex wall transitions are presented in Fig. 7.8. An image before and after the transition is presented for four wire widths. To guide the eye, vortex states are labeled with black and white arrows.

7.3.2 Results and Discussion

The temperature dependent transition as a function of ring width is shown in Fig. 7.9. In Fig. 7.9 (a), the temperature dependence of the number of domain walls that underwent a transition is shown. Extracted from these values, a mean

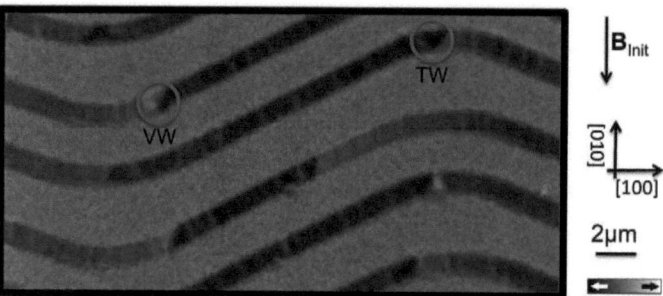

Figure 7.7: XMCD-PEEM image of magnetized Heusler alloy zig-zag wires. Domain walls separate the domains in the wire. Both vortex and transverse walls are present. In the straight segments, ripple domains are present. Some of the domain walls are not positioned at the kink.

switching field was determined and plotted against the ring width in Fig. 7.9 (b). A clear dependence on the ring width is observable. Note that this temperature dependence is not reflected in Fig. 7.8, as only selected examples are shown and the spread of the transition can be in the range of several hundred degrees (see Fig. 7.9 (a)).

Surprisingly, the graph in Fig. 7.9 (b) shows larger transition temperatures for wider elements. The more plausible result would have been a larger transition temperature for the narrower elements, as transverse walls are expected to be the stable domain spin configuration in narrow elements [58].

High temperatures lead to Cr-interdiffusion and re-crystallization. On four different ring widths, the drastic effect on the magnetic configuration and on the structure itself is presented in Fig. 7.10 for three different temperatures. It is observed, that the domains in the narrowest rings are destroyed at temperatures around 888 K. Before the ring is destroyed, small changes in the domain configuration can be observed. The remaining structures from 0.34 μm to 0.56μm wide elements show magnetic contrast at 888 K and also ripple domains that can be related to the structural changes. The complete thermal destruction of all elements is reached at around 898 K.

Even with these drastic structural changes, a spin modification at lower temperatures, where the structural changes have not set in, can be explained. The influence of these changes is seen to be more relevant at lower temperatures for narrower elements. If the domain wall transitions (Fig. 7.8) are caused or supported by the thermally activated crystallization and interdiffusion processes, then

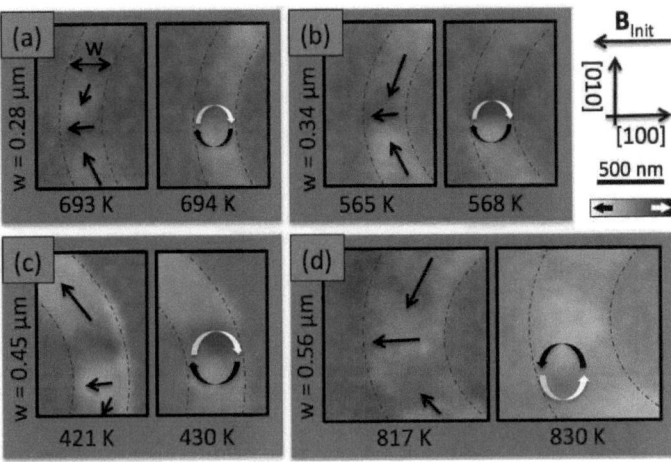

Figure 7.8: Examples of domain wall transitions in 30 nm thick Heusler alloy rings with widths from 280 nm (a) to 560 nm (d). For each ring, a transverse domain state prior (left) and a vortex domain state after (right) the transition is shown. The nucleated vortex is indicated by black and white arrows.

the slope of the graph in Fig. 7.9 can be explained, since smaller elements seem to be more affected.

7.4 Conclusion

Patterned Heusler alloy elements are fabricated on blank MgO and Cr seeded MgO substrates. The Cr seeded samples are ideally suited for PEEM experiments (no charging) where the static spin configurations can be mapped. To characterize the spin configuration in confined elements, the material was patterned with various geometries and sizes. The XMCD contrast of the elements decreases with increasing element size, which can be caused by size dependent anisotropy variation and/or annealing effects.

A small range of sizes with typical dimensions around 500 nm is observed, where the shape anisotropy dominates and domain walls in rings and wires can be generated reproducibly. This is particularly interesting for applications that rely on shape anisotropy-determined spin structures, such as domain wall based devices. Large elements show ripple domains and spin configurations that are less dominated by the particular element shape.

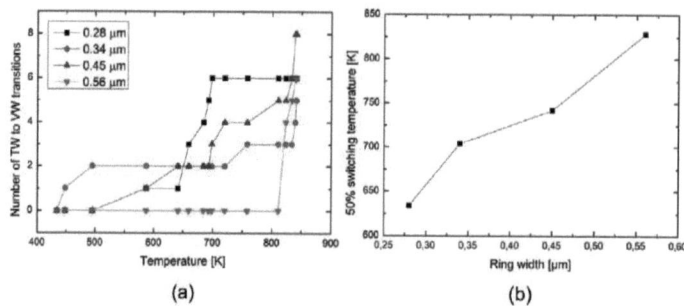

Figure 7.9: The number of temperature induced transverse wall to vortex wall transition in 30 nm thick Heusler alloy wires is shown in (a) for ring widths from 280 to 560 nm. The mean switching temperature over the ring width is presented in (b).

Heating experiments show the robustness of the spin configuration up to 800 K. Nevertheless, in this temperature range, Cr interdiffusion and crystalline changes cannot be neglected. At temperatures around 890 K, the patterned elements are physically destroyed. It is shown that the temperature of this permanent destruction is size dependent and narrow elements denature first.

Temperature and the change of the crystalline properties are observed to have a direct effect on the spin structure, such as the transition of transverse domain walls into vortex domain walls at temperatures between 400 K and 850 K. This is particularly important for devices, as it limits the usable temperature regime. However, the temperatures that are needed to induce permanent magnetic changes are well above room temperature.

As transverse walls are normally present in narrow rings, a higher transformation temperature for narrow rings would be expected in the temperature-induced domain wall changes. However, the opposite behavior was observed. This can be explained by a thermal depinning of the domain state that is supported by local crystalline changes. These crystalline changes occur at lower temperatures in narrow elements and therefore explain the observed trend.

In conclusion, the feasible control of the spin structure in patterned Heusler alloys ($Co_2FeAl_{0.4}Si_{0.6}$) is demonstrated. For a small size range around 500 nm, the spin structure is mainly determined by the element shape. The high spin polarization in these elements and the resistance to thermally activated changes make this material an interesting candidate for future applications and experiments. However, further optimization of the film quality and the patterning method to

7.4. Conclusion

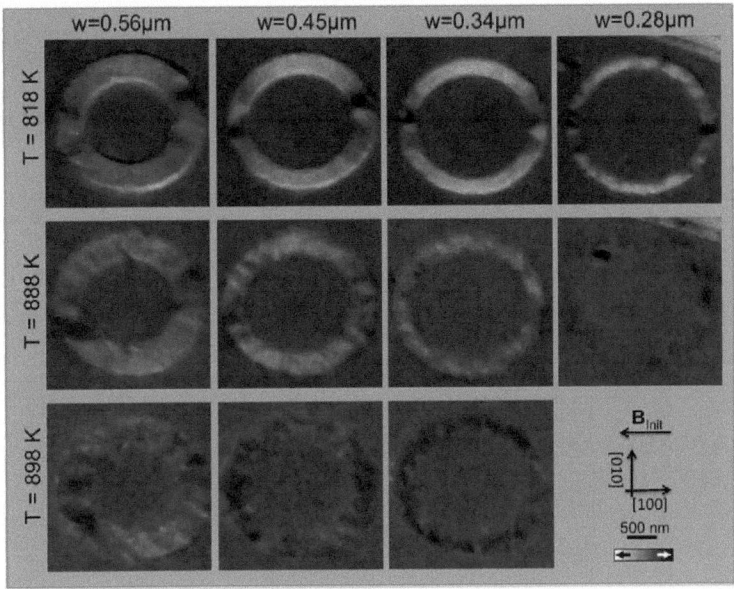

Figure 7.10: Temperature induced destruction of 30 nm thick Heusler alloy rings with different widths. The top row shows rings with magnetic contrast and no structural damages at 818 K. The center row shows the same rings at 888 K. In the three largest elements, small changes of the crystalline structure are visible but the magnetic contrast remains. The smallest structure, 0.28 μm wide, is destroyed. In the bottom row, at 898 K, the temperature induced changes also destroy the larger elements.

reduce local pinning are desirable.

CHAPTER 8

Conclusion and Outlook

This thesis focus on the investigation of spin structures and the field- and electric current-induced spin dynamics in micro- and nano-patterned magnetic elements in Permalloy and half-metallic materials.

Time resolved pump-probe experiments were carried out to study the field-induced domain wall dynamic in soft magnetic Permalloy (NiFe) nanowires. The unique possibilities available at the PEEM of the Swiss Light Source, a time resolution of 50 ps together with a spatial resolution of approximately 50 nm, allow for a unique insight into the first nanoseconds of the time response of the spin dynamics of an excited domain wall. Here, it was observed that the Zeeman energy of the field pulse deforms the spin structure prior the domain wall displacement, thus increasing the exchange energy, which was demonstrated by direct imaging. The delayed domain wall motion is followed by a fast domain wall displacement with velocities above 2000 m/s, which is not limited by the Walker breakdown due to the short time scale ($\sim$2-3 ns). After the field pulse decays, the domain wall relaxes back to its equilibrium position whilst undergoing a damped oscillation. These inertia-like effects, the delayed onset of domain wall motion and the oscillation around the equilibrium position, are based on the transformation of energy between different energy reservoirs, namely, the exchange and the Zeeman energy. Due to these quasi-classical characteristics, a domain wall mass of $(1.3 \pm 0.1) \times 10^{-24}$ kg could be determined for this quasi-particle.

The dynamics of current-induced domain wall motion in Permalloy was studied with a view towards understanding and minimizing pinning processes that hinder domain wall dynamics in confined nanostructures. The key for improving the current-induced domain wall motion was the development of a special pattern transfer method that uses a negative electron beam lithography resist and Ar-ion milling to reduce edge roughness and edge pinning. By measuring current-induced domain wall motion using magnetoresistance response of the system, we find that the wires prepared with this new lithography technique exhibit critical current densities that are about four times smaller than those obtained in conventionally patterned wires. This demonstrates that controlling pinning is key for reducing the critical current density.

To further improve the spin-torque efficiency, materials that show both a low saturation magnetization and a high spin polarization are highly desirable. Ferromagnetic half-metals are promising candidates, as they can provide both properties. This class of material needs special growth conditions and annealing steps, which requires adapted patterning methods. Also magnetocrystalline anisotropies in these epitaxial systems are more dominant than in, e.g., polycrystalline Permalloy. As a first step towards investigating the suitability of these materials for spin-torque applications the spin configurations in confined structures of two different half-metallic systems, LSMO ($La_{0.7}Sr_{0.3}MnO_3$) and a Heusler alloy ($Co_2FeAl_{0.4}Si_{0.6}$), were investigated.

In $La_{0.7}Sr_{0.3}MnO_3$ (LSMO) elements, the dimensional crossover from multidomain to shape-defined magnetic states in highly spin-polarized LSMO structures was determined. Weak pinning and low magnetic anisotropies give rise to highly symmetric states that are determined by the shape anisotropy. In particular, it was found that well defined domain walls are generated in spatially confined geometries such as rings, whose character can be controlled by varying the width of the element. In addition, it is demonstrated that the low energy states are robust against thermal excitation up to the critical temperature. These results show that LSMO is a promising candidate for both the study of fundamental domain wall phenomena in highly spin polarized materials and for device applications, where robust and well determined spin configurations are key.

Another class of ferromagnetic half-metals are the Heusler alloys. A lift-off technique was developed to manufacture patterned elements of $Co_2FeAl_{0.4}Si_{0.6}$. In rings and nanowires, transverse walls that are predominant after magnetization are found to transform into a vortex wall spin structure at temperatures around

~623 K. For a judicious choice of dimensions flux closure spin structures, and well defined domain walls, are observed. This is a prerequisite for future spin dynamic experiments and applications, since the work principle is based on the reliability of the domain wall spin structure.

Magnetization dynamics, induced by field and spin currents, are of high interest for science and industry. The results of the time resolved spin dynamics show that the domain wall mass influences strongly the domain wall dynamics in the first hundreds of picoseconds, and therefore cannot be neglected in high performance devices that rely on domain wall dynamics. In device applications, were a fast response to an applied pulse is desirable, a smaller domain wall mass may improve the performance, which could be achieved by varying the size and/or the type of the domain wall.

Possible candidates to improve current-induced domain wall motion are half-metallic ferromagnets like LSMO, and Heusler alloys with a theoretically predicted spin polarization up to 100%. In addition to an improvement of the performance of the current-induced domain wall motion, these materials are ideally suited for spin injectors in spintronic applications and experiments. The gained information about the spin configuration in these novel half-metallic materials pave the way for future experiments based on the spin dynamics induced by highly spin polarized currents. The expected improvement of the spin-torque efficiency may be sufficient to reduce the required critical current density, which is a prerequisite for device applications based on current-induced domain wall displacement.

Bibliography

[1] G. E. Moore, *IEEE Text Speech* (1975).

[2] M. N. Baibich, et al., *Phys. Rev. Lett.* **61**, 2472 (1988).

[3] G. Binasch, P. Grünberg, F. Saurenbach, W. Zinn, *Phys. Rev. B* **39**, 4828 (1989).

[4] A. Moser, et al., *J. Phys. D: Appl. Phys.* **35**, R157 (2002).

[5] H. Zhou, H. N. Bertram, *IEEE* **35**, 2712 (1999).

[6] R. Cowburn, *U.S. Patent No. WO/2007/132174* (2007).

[7] S. S. P. Parkin, M. Hayashi, L. Thomas, *Science* **320**, 190 (2008).

[8] J. Rhensius, et al., *Phys. Rev. Lett.* **104**, 067201 (2010).

[9] G. Malinowski, et al., *J. Phys. D: Appl. Phys.* **43**, 45003 (2010).

[10] H. Hidaka, *Embedded Memories for Nano-Scale VLSIs*, K. Zhang, ed., Series on Integrated Circuits and Systems (Springer-Verlag, 2009), p. 241.

[11] J. Stöhr, H. C. Siegmann, *Magnetism from Fundamentals to Nanoscale Dynamics* (Springer, 2006).

[12] H. Ibach, H. Lüth, *Festkörperphysik* (Springer, Berlin Heidelberg New York, 1999).

[13] C. Zener, *Phys. Rev.* **82**, 403 (1951).

[14] Y. Tokura, Y. Tomioka, *J. Magn. Magn. Mater.* **200**, 1 (1999).

[15] J. W. F. Brown, *Micromagnetics* (Krieger, New York, 1978).

[16] A. Aharoni, *Introduction to the Theory of Ferromagnetism* (Oxford Science publication, 2000).

[17] http://math.nist.gov/oommf/.

[18] A. Hubert, R. Schäfer, *Magnetic domains. The Analysis of Magnetic Microstructures* (Springer Verlag, 1998).

[19] B. Heinrich, *Ultrathin magnetic structures III*, J. Bland, B. Heinrich, eds. (Springer-Verlag, Berlin Heidelberg, 2005), p. 143.

[20] J. Slonczewski, *Magn. Magn. Mater.* **159**, L1 (1996).

[21] A. Thiaville, Y. Nakatani, J. Miltat, N. Vernier, *J. Appl. Phys.* **95**, 7049 (2004).

[22] S. Zhang, S. Li, *Phys. Rev. Lett.* **93**, 127204 (2004).

[23] Z. Li, S. Zhang, *Phys. Rev. B* **70**, 024417 (2004).

[24] W. Thomson, *Proc. Roy. Soc. London* **8**, 546 (1857).

[25] L. Heyne, et al., *Appl. Phys. Lett.* **96**, 32504 (2010).

[26] L. Heyne, et al., *Phys. Rev. B* **80**, 184405 (2009).

[27] P. Möhrke, J. Rhensius, J.-U. Thiele, L. J. Heyderman, M. Kläui, *Solid State Communications* **150**, 489 (2009).

[28] L. Heyne, M. Kläui, J. Rhensius, L. L. Guyader, F. Nolting, *Review of Scientific Instruments* **81**, 113707 (2010).

[29] D. Ilgaz, et al., *Phys. Rev. Lett.* **105**, 76601 (2010).

[30] L. Heyne, et al., *Phys. Rev. Lett.* **105**, 187203 (2010).

[31] M. Eltschka, et al., *Phys. Rev. Lett.* **105**, 056601 (2010).

[32] A. K. Patra, et al., *Phys. Rev. B* **82**, 134447 (2010).

[33] T. A. Moore, et al., *J. Magn. Magn. Mater.* **322**, 1347 (2010).

[34] J. H. Franken, et al., *Appl. Phys. Lett.* **95**, 212502 (2009).

[35] A. Bisig, et al., *Appl. Phys. Lett.* **96**, 152506 (2010).

[36] T. Moore, et al., *Phys. Rev. B* **80**, 132403 (2009).

[37] C. Moutafis, et al., *Submitted* (2011).

[38] J. Rhensius, et al., *Appl. Phys. Lett.* **99**, 062508 (2011).

[39] A. Bieren, et al., *Submitted* (2011).

[40] http://www.cnf.cornell.edu/image/spiefig20.jpg.

[41] A. Aziz, et al., *J. Appl. Phys.* **98**, 124102 (2005).

[42] J. L. Keddie, R. A. L. Jones, R. A. Cory, *Faraday Discuss.* **98** (1994).

[43] D. Backes, Spin structure of domainwalls and their behaviour in applied fields and currents, Master's thesis, University of Konstanz (2008).

[44] H. Namatsu, T. Yamaguchi, M. Nagase, K. Yamazaki, K. Kurihara, *Microelectron. Eng.* **42**, 331 (1998).

[45] P. Perna, et al., *arXiv:1005.0553v3 [cond-mat.mes-hall]* (2010).

[46] C. A. F. Vaz, Y. Segal, J. Hoffman, F. J. Walker, C. H. Ahn, *J. Vac. Sci. Technol. B* **28**, C5A6 (2010).

[47] R. Cowburn, *U.S. Patent No. WO/2007/132174* (2007).

[48] W. Döring, *Z. Naturforsch* **3**, 373 (1948).

[49] E. Saitoh, H. Miyajima, T. Yamaoka, , G. Tatara, *Nature* **432**, 203 (2004).

[50] D. Bedau, et al., *Phys. Rev. Lett.* **99**, 146601 (2007).

[51] J. Raabe, et al., *Phys. Rev. Lett.* **94**, 217204 (2005).

[52] M. Hayashi, et al., *Nature Phys.* **3**, 21 (2007).

[53] M. Hayashi, et al., *Science* **320**, 209 (2008).

[54] T. Ono, et al., *Science* **284**, 468 (1999).

[55] D. Atkinson, et al., *Nature Mater.* **2**, 85 (2003).

[56] G. S. D. Beach, et al., *Nature Mater.* **4**, 741 (2005).

[57] D. Backes, et al., *Appl. Phys. Lett.* **91**, 112502 (2007).

[58] M. Kläui, et al., *Appl. Phys. Lett.* **85**, 5637 (2004).

[59] A. Yamaguchi, et al., *Phys. Rev. Lett.* **92**, 077205 (2004).

[60] L. J. Heyderman, et al., *Microelectron. Eng.* **73-74**, 780 (2004).

[61] O. Boulle, et al., *Journal of Applied Physics* **105**, 07C106 (2009).

[62] A. Thiaville, Y. N. et al., *Spin Dynamics in Confined Magnetic Structures III and edited by B. Hillebrands* (Springer and New York, 2003).

[63] C. Kittel, *Phys. Rev.* **80**, 918 (1950).

[64] L. Berger, *Phys. Rev. B* **33**, 1572 (1986).

[65] J. Grollier, et al., *Appl. Phys. Lett.* **83**, 509 (2003).

[66] N. Vernier, D. A. Allwood, D. Atkinson, M. D. Cooke, R. P. Cowburn, *Europhys. Lett.* **65**, 526 (2004).

[67] M. Kläui, et al., *Phys. Rev. Lett.* **95**, 026601 (2005).

[68] A. Himeno, S. Kasai, T. Ono, *Appl. Phys. Lett.* **87**, 243108 (2005).

[69] R. J. Soulen Jr., et al., *Science* **282**, 85 (1998).

[70] F. Junginger, et al., *Appl. Phys. Lett.* **90**, 132506 (2007).

[71] E. M. Hempe, et al., *Phys. Status Solidi a* **204**, 3922 (2007).

[72] M. Hayashi, et al., *Phys. Rev. Lett.* **97**, 207205 (2006).

[73] P. Möhrke, et al., *J. Phys. D:* **41**, 164009 (2008).

[74] M. Laufenberg, et al., *Phys. Rev. Lett.* **97**, 046602 (2006).

[75] J. Heinen, et al., *Appl. Phys. Lett.* **96**, 202510 (2010).

[76] M. Kläui, et al., *Appl. Phys. Lett.* **88**, 232507 (2006).

[77] L. Thomas, S. S. P. Parkin, *Handbook of Magnetism and Magnetic Materials* **3**, 942 (2007).

[78] J. Coey, M. Venkatesan, *J. Appl. Phys.* **91**, 8345 (2002).

[79] J.-H. Park, et al., *Nature* **392**, 794 (1998).

[80] G. H. Jonker, J. H. van Santen, *Physica* **16**, 337 (1950).

[81] C. A. F. Vaz, C. H. Ahn, V. E. Henrich, *Epitaxial ferromagnetic films and spintronic applications*, A. Hirohata, Y. Otani, eds. (Research Signpost, 2009), p. 145.

[82] J. Cibert, J.-F. Bobo, U. Lüders, *C. R. Physique* **6**, 977 (2005).

[83] H. J. A. Molegraaf, et al., *Adv. Mater.* **21**, 3470 (2009).

[84] C. A. F. Vaz, et al., *Phys. Rev. Lett.* **104**, 127202 (2010).

[85] X. Hong, A. Posadas, A. Lin, C. H. Ahn, *Phys. Rev. B* **68**, 134415 (2003).

[86] Y. Wu, Y. Matsushita, Y. Suzuki, *Phys. Rev. B* **64**, 220404(R) (2001).

[87] A. Biehler, et al., *Phys. Rev. B* **75**, 184427 (2007).

[88] H. S. Goripati, et al., *J. Appl. Phys.* **109**, 043901 (2011).

[89] C. Quitmann, et al., *Nuclear Instruments and Methods in Physics Research A* **588**, 494 (2008).

[90] M. Laufenberg, et al., *Appl. Phys. Lett.* **88**, 052507 (2006).

[91] J. Rothman, et al., *Phys. Rev. Lett.* **86**, 1098 (2001).

[92] M. Kläui, et al., *J. Magn. Magn. Mater.* **272**, 1631 (2004).

[93] M. Kläui, et al., *J. Magn. Magn. Mater.* **240**, 7 (2002).

[94] M. Kläui, C. A. F. Vaz, J. A. C. Bland, L. J. Heyderman, *Appl. Phys. Lett.* **86** (2005).

[95] Y. G. Yoo, M. Kläui, C. A. F. Vaz, L. J. Heyderman, J. A. C. Bland, *Appl. Phys. Lett.* **82**, 2470 (2003).

[96] L. Lopez-Diaz, J. Rothman, M. Klani, J. Bland, *IEEE* **36**, 3155 (2000).

[97] A. Adeyeye, et al., *J. Appl. Phys.* **79**, 6120 (1996).

[98] A. Adeyeye, J. Bland, C. Daboo, D. Hasko, *Phys. Rev. Lett.* **56**, 3265 (1997).

[99] Y. Takamura, et al., *Nano Letters* **6**, 1287 (2006).

[100] M. Kubota, et al., *Appl. Phys. Lett.* **91**, 182503 (2007).

[101] Y. Takamura, R. Nakane, H. Munekata, S. Sugahara, *J. Appl. Phys.* **103**, 07D719 (2008).

[102] I. Galanakis, P. H. Dederichs, N. Papanikolaou, *Phys. Rev. B* **66**, 174429 (2002).

[103] N. Tezuka, et al., *Appl. Phys. Lett.* **89**, 112514 (2006).

[104] S. Wurmehl, G. Fecher, H. Kandpal, V. Ksenofontov, C. Felser, *Appl. Phys. Lett.* **88**, 032503 (2006).

[105] R. A. de Groot, F. M. Mueller, P. G. v. Engen, K. H. J. Buschow, *Phys. Rev. Lett.* **50**, 2024 (1983).

[106] K. H. Bushow, P. G. v. Engen, R. Jongebreur, *J. Magn. Magn. Mater.* **38**, 1 (1983).

[107] J. Heinen, J. Rhensius, M. Kläui, T. Graf, C. Felser, *DGP Frühjahrstagung* **MA 63.61** (2011).

[108] M. Fonin, et al., *J. Appl. Phys.* **109**, 07D351 (2011).

Publication List

- S. Bedanta, T. Eimüller, W. Kleemann, J. Rhensius, F. Stromberg, E. Amaladass, S. Cardoso, and P. P. Freitas. Overcoming the Dipolar Disorder in Dense CoFe Nanoparticle Ensembles: Superferromagnetism, *Phys. Rev. Lett.* **98**, 176601 (2007).

- W. Kleemann, J. Rhensius, O. Petracic, J. Ferré, J. P. Jamet, and H. Bernas. Modes of Periodic Domain Wall motion in Ultrathin Ferromagnetic Layers, *Phys. Rev. Lett.* **99**, 97203 (2007).

- S. Bedanta, J. Rhensius, W. Kleemann, P. Parashar, S. Cardoso, and P. P. Freitas. Dynamic magnetization properties of a superferromagnetic metal-insulator multilayer observed by magneto-optic Kerr microscopy, *Journal of Applied Physics* **105**, 07C306 (2009).

- O. Boulle, L. Heyne, J. Rhensius, M. Kläui, U. Rüdiger, L. Joly, L. Le Guyader, F. Nolting, L. J. Heyderman, G. Malinowski, H. J. M. Swagten, B. Koopmans, C. Ulysse, and G. Faini. Reversible switching between bidomain states by injection of current pulses in a magnetic wire with out-of-plane magnetization, *Journal of Applied Physics* **105**, 07C106 (2009).

- J. H. Franken, P. Möhrke, M. Kläui, J. Rhensius, L. J. Heyderman, J.-U. Thiele, H. J. M. Swagten, U. J. Gibson, and U. Rüdiger. Effects of combined current injection and laser irradiation on Permalloy microwire switching, *Appl. Phys. Lett.* **95**, 212502 (2009).

- L. Heyne, J. Rhensius, Y.-J. Cho, D. Bedau, S. Krzyk, C. Dette, H. S. Körner, J. Fischer, M. Laufenberg, D. Backes, L. J. Heyderman, L. Joly, F. Nolting, G. Tatara, H. Kohno, S. Seo, U. Rüdiger, and M. Kläui. Geometry-dependent scaling of critical current densities for current-induced domain wall motion and transformations, *Phys. Rev. B* **80**, 184405 (2009).

- P. Möhrke, J. Rhensius, J.-U. Thiele, L. J. Heyderman, and M. Kläui. Tailoring laser-induced domain wall pinning, *Solid State Communications* **150**, 489 (2009).

- L. Heyne, J. Rhensius, A. Bisig, S. Krzyk, P. Punke, M. Kläui, L. J. Heyderman, L. Le Guyader, and F. Nolting. Direct observation of high velocity current induced domain wall motion, *Appl. Phys. Lett.* **96**, 32504 (2010).

- G. Malinowski, A. Lörincz, S. Krzyk, P. Möhrke, D. Bedau, O. Boulle, J. Rhensius, L. J. Heyderman, Y. J. Cho, S. Seo, and M. Kläui. Current-induced domain wall motion in $Ni_{80}Fe_{20}$ nanowires with low depinning fields, *J. Phys. D: Appl. Phys.* **43**, 45003 (2010).

- T. A. Moore, M. Kläui, P. Möhrke, D. Backes, J. Rhensius, U. Rüdiger, L. J. Heyderman, J.-U. Thiele, G. Woltersdorf, C.H. Back, A. Fraile Rodríguez, F. Nolting, T.O. Mentes, M. Á. Niño, A. Locatelli, A. Potenza, H. Marchetto, S. Cavill, and S.S. Dhesi. Scaling of spin relaxation and angular momentum dissipation in permalloy nanowire. *Phys. Rev. B* **80**, 132403 (2009).

- J. Rhensius, L. Heyne, D. Backes, S. Krzyk, L. J. Heyderman, L. Joly, F. Nolting, and M. Kläui. Imaging of Domain Wall Inertia in Permalloy Half-Ring Nanowires by Time-Resolved Photoemission Electron Microscopy, *Phys. Rev. Lett.* **104**, 067201 (2010).

- A. Bisig, J. Rhensius, M. Kammerer, M. Curcic, H. Stoll, G. Schütz, B. Van Waeyenberge, K. Wei Chou, T. Tyliszczak, L. J. Heyderman, S. Krzyk, A. v. Bieren, and M. Kläui. Direct imaging of current induced magnetic vortex gyration in an asymmetric potential well, *Appl. Phys. Lett.* **96**, 152506 (2010).

- T. A. Moore, M. Kläui, L. Heyne, P. Möhrke, D. Backes, J. Rhensius, U. Rüdiger, L. J. Heyderman, T. O. Mentes, M. Á. Niño, A. Locatelli, A. Potenza, H. Marchetto, S. Cavild, and S. S. Dhesi. Domain wall velocity measurement in permalloy nanowires with x-ray magnetic circular dichroism imaging and single shot kerrmicroscopy, *Journal of Magnetism and Magnetic Materials* **322**, 1347 (2010).

- M. Eltschka, M. Wötzel, J. Rhensius, S. Krzyk, U. Nowak, M. Kläui, T. Kasama, R. E. Dunin-Borkowski, L. J. Heyderman, H. J. van Driel, and R. A. Duine. Nonadiabatic spin torque investigated using thermally activated magnetic domain wall dynamics, *Phys. Rev. Lett.* **105**, 056601 (2010).

- D. Ilgaz, J. Nievendick, L. Heyne, D. Backes, J. Rhensius, T. A. Moore, M. Á. Niño, A. Locatelli, T. O. Mentes, A. v. Schmidsfeld, A. v. Bieren, S. Krzyk, L. J. Heyderman, and M. Kläui. Domain-wall depinning assisted by pure spin currents, *Phys. Rev. Lett.* **105**, 76601 (2010).

- A. K. Patra, A. von Bieren, S. Krzyk, J. Rhensius, L. J. Heyderman, R. Hoffmann, and M. Kläui. Magnetoresistance measurement of tailored permalloy nanocontacts, *Phys. Rev. B* **82**, 134447 (2010).

- L. Heyne, J. Rhensius, D. Ilgaz, A. Bisig, U. Rüdiger, M. Kläui, L. Joly, F. Nolting, L. J. Heyderman, J. U. Thiele, and F. Kronast. Direct Determination of Large Spin-Torque Nonadiabaticity in Vortex Core Dynamics, *Phys. Rev. Lett.* **105**, 187203 (2010).

- L. Heyne, M. Kläui, J. Rhensius, L. Le Guyader, and F. Nolting. In situ contacting and current-injection into samples in photoemission electron microscopes, *Review of Scientific Instruments* **81**, 113707 (2010).

- J. Rhensius, C. A. F. Vaz, A. Bisig, S. Schweitzer, J. Heidler, H. S. Körner, A. Locatelli, M. Á. Niño, M. Weigand, L. Méchin, F. Gaucher, E. Göring, L. J. Heyderman, and M. Kläui. Control of spin configuration in half-metallic $La_{0.7}Sr_{0.3}MnO_3$ nano-structures, *Appl. Phys. Lett.* **99**, 062508 (2011).

- A. C. von Bieren, A. K. Patra, S. Krzyk, J. Rhensius, L. J. Heyderman, R. Hoffmann, and M. Kläui. Domain-wall induced large magnetoresistance effects at zero applied field in electromigrated Permalloy nanocontacts, *Submitted* (2011).

- C. Moutafis, J. Rhensius, A. Bisig, F. Büttner, C. Barton, C. Morrison, T. Thomson, C. Tieg, S. Schaffert, B. Pfau, C. Günther, S. Eisebitt, and M. Kläui. Skyrmions in Perpendicular Magnetic Anisotropy Dots: Imaging and Simulations, *Submitted* (2011).

i want morebooks!

Buy your books fast and straightforward online - at one of world's fastest growing online book stores! Environmentally sound due to Print-on-Demand technologies.

Buy your books online at
www.get-morebooks.com

Kaufen Sie Ihre Bücher schnell und unkompliziert online – auf einer der am schnellsten wachsenden Buchhandelsplattformen weltweit! Dank Print-On-Demand umwelt- und ressourcenschonend produziert.

Bücher schneller online kaufen
www.morebooks.de

VDM Verlagsservicegesellschaft mbH
Heinrich-Böcking-Str. 6-8   Telefon: +49 681 3720 174   info@vdm-vsg.de
D - 66121 Saarbrücken   Telefax: +49 681 3720 1749   www.vdm-vsg.de

Printed by Books on Demand GmbH, Norderstedt / Germany